Musket Ball and Small Shot Identification

Musket Ball and Small Shot Identification

A Guide

Daniel M. Sivilich

Foreword by David Gerald Orr
Introduction by Douglas D. Scott
Appendix by Henry M. Miller

University of Oklahoma Press
Norman

Credit lines for images used as design elements in this book:
page 47, reenactment photographs by Patricia Robinson
page 92, reenactment photograph by Michael Smith
page 102, Wilde Boar, woodcut by Edward Topsell, 1658. (Image courtesy of Special Collections, University of Houston Libraries, University of Houston Digital Library.)
Credit lines for cover images are on the colophon page.

Library of Congress Cataloging-in-Publication Data

Sivilich, Daniel M., 1950– author.
    Musket ball and small shot identification : a guide / Daniel M. Sivilich ; foreword by David Gerald Orr ; introduction by Douglas D. Scott ; Appendix B by Henry M. Miller.
        pages cm
    Includes bibliographical references and index.
    ISBN 978-0-8061-5158-8 (pbk. : alk. paper)
1. Bullets—Identification. 2. Shot (Pellets)—Identification. 3. Bullets—History—18th century. 4. Firearms—United States—History—18th century. 5. United States—History—Revolution, 1775–1783—Equipment and supplies. 6. United States—History—Revolution, 1775–1783—Antiquities. 7. Monmouth Battlefield State Park (N.J.)—Antiquities. I. Title.

UF770.S53 2016
623.4'55027—dc23

2015032059

The paper in this book meets the guidelines for permanence and durability of the Committee on Production Guidelines for Book Longevity of the Council on Library Resources, Inc. ∞

Copyright © 2016 by the University of Oklahoma Press, Norman, Publishing Division of the University. Manufactured in the U.S.A.

All rights reserved. No part of this publication may be reproduced, stored in a retrieval system, or transmitted, in any form or by any means, electronic, mechanical, photocopying, recording, or otherwise—except as permitted under Section 107 or 108 of the United States Copyright Act—without the prior written permission of the University of Oklahoma Press. To request permission to reproduce selections from this book, write to Permissions, University of Oklahoma Press, 2800 Venture Drive, Norman, OK 73069, or email rights.oupress@ou.edu.

1 2 3 4 5 6 7 8 9 10

To the many members of the Battlefield Restoration and Archaeological Volunteer Organization (BRAVO)

Contents

List of Illustrations   ix

List of Tables   xvii

Foreword, *by David Gerald Orr*   xix

Preface   xxi

Acknowledgments   xxiii

Introduction, *by Douglas D. Scott*   xxix

1   The Gun and Early Projectiles   3

2   The Basic Musket Ball   16

3   What Did It Hit?   47

4   Musket Balls with Fabric Impressions   66

5   Musket Balls Altered to Improve Lethality   73

6   Canister Shot   92

7   Chewed Musket Balls   102

8   "Pewter" Musket Balls   116

9   Musket Balls Altered for Nonlethal Use   128

10   Small Shot . . . Not Just for the Birds   145

Epilogue   153

Appendix A The Chronology of the Early Gun **155**

Appendix B An Analysis of Dentition Marks on Musket Balls,
by Henry M. Miller **159**

Appendix C Classifications of Small Lead Shot **170**

Appendix D BRAVO **172**

References **177**

Index **187**

Illustrations

Figures

1.1	*Landing of Columbus*, 1836 (Vanderlyn)	3
1.2	Roman and Greek sling shots	4
1.3	Seventeenth-century lead-covered stone musket ball from Ukraine	5
1.4	Seventeenth-century glass musket ball from Poland	5
1.5	Earliest known picture of a gun	6
1.6	Two "handgonnes" found in Switzerland	6
1.7	Reproduction matchlock musket	7
1.8	Reproduction wheel-lock musket	7
1.9	Early seventeenth-century colonial snaphaunce lock	8
1.10	Model 1727 long land pattern British Brown Bess musket	8
1.11	Tower-marked Model 1777 military British Brown Bess musket	9
1.12	Model 1768 French military musket	9
1.13	Musket lock excavated at Valley Forge	11
1.14	Close-up of excavated musket lock	12
1.15	X-ray of excavated musket lock	12
1.16	Cleaned musket lock	13
1.17	Reproduction French Charleville lock	13
1.18	Model 1728 French musket	14
1.19	Cache of bayonets excavated at Valley Forge	14
1.20	Bayonets excavated at Valley Forge	15
2.1	Artifacts from 1655–75 Seneca village Dann site	16
2.2	Steatite musket ball and small shot mold	17

ILLUSTRATIONS

2.3	Musket ball with mold seam and casting sprue visible	19
2.4	Musket ball with unclipped sprue and lead waste from a gang mold	19
2.5	Reproduction military cartridge	20
2.6	Diameter distribution of musket shot at Jamestown	21
2.7	Musket balls excavated at Valley Forge, Fort Montgomery, and New York City	22
2.8	Musket balls with different diameters, Monmouth Battlefield State Park	23
2.9	Musket ball with air pocket visible, Monmouth Battlefield State Park	25
2.10	Calculated musket ball diameters using density value of pure lead vs. Sivilich Formula	27
2.11	Analysis of musket balls from American camp site, New Jersey	28
2.12	Analysis of musket balls from British-occupied site, New Jersey	30
2.13	Analysis of musket balls excavated at Monmouth Battlefield State Park	32
2.14	Three musket balls with buckshot impressions	33
2.15	Typical example of buck and ball	33
2.16	Musket ball and three-buckshot packing diagram	34
2.17	Buck and ball excavated as a cluster	34
2.18	Size distribution of buck and ball cluster	35
2.19	Pre–Civil War and Civil War–era buck and ball	35
2.20	Musket balls with ramrod marks	36
2.21	Reproduction ramrods	37
2.22	Musket balls extracted with a screw, excavated at Battle of Pułtusk, Poland	37
2.23	Musket balls extracted with a screw, excavated at American camp sites	38
2.24	British musket worm excavated at 1777 Saratoga Battlefield	39
2.25	Musket balls extracted with worms from American camp site, New Jersey	39
2.26	Musket worm	40
2.27	Musket balls with worm marks and buckshot impressions	40
2.28	Rejected musket balls	41
2.29	Musket balls from Battle of Boyne and near Ballymore, Ireland, 1690s	42

2.30	Austrian/Hungarian military musket cartridge and reproduction paper cartridge	42
2.31	Lead musket ordnance excavated at 1649 Battle of Zboriv, Ukraine	43
2.32	Lead musket ordnance excavated at 1677 Battle of Landskrona, Sweden	43
2.33	Musket ball with extended sprue and reproduction lead ball with casting sprue	44
2.34	Lead shot used by matchlock-type muskets and pistols	44
2.35	Jug-shaped musket shot from Battle of Poltava, Poland, and Cheshire, England	45
2.36	Artifacts excavated at 1655–75 Seneca village sites, New York	45
2.37	Extended sprue musket balls, Seneca village Dann site	46
3.1	Musket balls that struck soft materials	48
3.2	Musket balls with "barrel bands"	49
3.3	Reproduction musket ball fired into a tree	50
3.4	Reproduction musket ball bisections	50
3.5	Musket balls that appear to have hit trees	51
3.6	Reproduction musket balls that hit trees	52
3.7	Musket ball that lodged in a tree	52
3.8	Musket ball that lodged in a tree and was hit by smaller projectiles	53
3.9	Musket balls that ricocheted off solid objects	53
3.10	Musket balls that ricocheted off rocks	54
3.11	Musket ball that hit a fence	55
3.12	Musket ball that hit a smooth, hard, flat object	55
3.13	Musket balls that probably hit musket barrels	56
3.14	Musket balls embedded in human bones	58
3.15	Musket ball with front tooth impression	58
3.16	Surgically extracted large-diameter musket ball and reproduction extractor	59
3.17	Fired musket ball with extractor screw hole	60
3.18	Reproduction musket balls struck with reproduction ramrod	61
3.19	Fired reproduction musket ball with extractor screw hole	61
3.20	Musket ball with deep impression, may have hit sword guard	62
3.21	Musket ball with deep, narrow impression, may have hit sword blade	63
3.22	Two musket balls fused together by midair collision	64

ILLUSTRATIONS

4.1	Musket ball and buckshot molds	66
4.2	Musket balls with fabric wadding marks	67
4.3	Musket ball with fabric patch impression and rifling marks	69
4.4	Musket ball with fabric patch impression	70
4.5	Lightly impacted musket ball with fabric patch impression	70
4.6	Impacted large-diameter ball with fabric patch impression	71
5.1	Halved musket balls	74
5.2	Quartered musket balls	75
5.3	Two views of quartered musket ball	76
5.4	Musket balls with nails	78
5.5	Cylindrical shot excavated at Monmouth Battlefield State Park	79
5.6	Cylindrical shot of similar sizes and shapes	79
5.7	Cylindrical shot recovered from pirate ship *Whydah*	82
5.8	Musket balls hammered into cylindrical shot	83
5.9	Cylindrical shot excavated at 1691 Battle of Aughrim, Ireland	83
5.10	Cylindrical shot excavated at 1714 Talamanca Battlefield, Spain	84
5.11	Cylindrical shot found at American camp site, New Jersey	85
5.12	Two musket balls made into "sluggs" excavated on Pułtusk Battlefield, Poland	85
5.13	Cylinder shot from Coronado expedition into New Mexico	85
5.14	Fabric impressions in cylinder shot	85
5.15	Musket ball with cavity and clay casting of cavity	87
5.16	Musket balls with possible casting flaw pinholes	87
5.17	Musket balls with possible casting flaws	88
5.18	Round musket ball with wire through	89
5.19	Double shot musket ball cast or fused together	89
5.20	Model 1768 French military musket cock with leather flint wrap	90
5.21	Flint with lead wrap	90
5.22	Flint wraps made from rifle balls	91
5.23	Worn flint wrap	91
5.24	Worn flint wrap with jaw serrations	91
6.1	Fused musket balls	92
6.2	Three pairs of fused musket balls	92

6.3	Musket balls used as canister shot	93
6.4	Canister shot containers	93
6.5	Early nineteenth-century grapeshot for four-pounder cannon	94
6.6	Canister shot with different shapes	96
6.7	Shot from tin canister and original packing arrangement	97
6.8	Lead-alloy hardness scale	98
6.9	Aerial photograph showing canister shot, Sutfin orchard, Monmouth Battlefield	99
6.10	Canister shot excavated from 1759 Battle of Kunersdorf, Poland	100
6.11	Modern shrapnel balls found with eighteenth-century military artifacts, West Point	101
7.1	Severely mashed, swine-chewed musket balls	103
7.2	Swine-chewed musket balls with incisor depressions	104
7.3	Swine-chewed musket balls with shallow dentition marks	104
7.4	Swine-chewed musket ball with lead loss	104
7.5	Swine-chewed musket ball	105
7.6	Swine molar fragment with lead embedded	105
7.7	Rat damage to a section of garden hose	106
7.8	Rodent damage to lead die	106
7.9	Impacted musket ball chewed by large rodent	106
7.10	Musket balls chewed by small rodents	107
7.11	Musket ball with curved bite marks and replication of marks using deer jaw	108
7.12	Reproduction musket balls chewed with human incisors, canines, and molars	110
7.13	Reproduction lead musket ball chewed with human canines and molars	111
7.14	Twelve reproduction musket balls and the mold used to cast them	111
7.15	Eighteenth-century musket ball and reproduction chewed with human molars	112
7.16	Seventeenth-century round musket ball and human-chewed musket ball	112
7.17	Eighteenth-century musket balls and reproductions chewed with human canines	113
7.18	Musket ball excavated at West Point and reproduction chewed with human canines	114

ILLUSTRATIONS

8.1	Musket ball diameter versus weight curves of lead/tin alloys	118
8.2	"Pewter" alloy musket balls, Monmouth Battlefield State Park	118
8.3	"Pewter" alloy American regimental buttons, Valley Forge	119
8.4	"Pewter" alloy gaming die, Valley Forge	120
8.5	Lead and "pewter" alloy musket balls, Battle of Cooch's Bridge, Delaware	120
8.6	Impacted "pewter" alloy musket balls, Fort Montgomery, New York	121
8.7	*The Destruction of the Royal Statue in New York*, ca. 1776	121
8.8	Fragments of gilded, leadened statue of King George III	123
8.9	Statue fragment with cut marks and a pick hole	123
8.10	Michael Seibert testing statue fragment using X-ray fluorescence spectroscopy	123
8.11	Spectrum of exterior surface of statue fragments	124
8.12	Spectrum of exterior and interior surfaces of statue fragments	124
8.13	Iron wire frame in reverse side of statue fragment	125
8.14	Iron support bar inside statue fragment	125
8.15	Tin spectrum for musket ball compared to statue fragments	126
8.16	Pewter button with raised "USA/1777"	127
9.1	May 8, 1777, general order from Washington	129
9.2	May 26, 1777, general order from Washington	130
9.3	Gaming dice made from musket balls, Valley Forge	131
9.4	Reproductions showing transformation from musket ball to gaming die	131
9.5	Unfinished gaming dice made from musket balls, Valley Forge	132
9.6	January 28, 1778, general order from Washington	133
9.7	Possible unfinished gaming dice made from musket balls, Valley Forge	134
9.8	Sketches of lead gaming tokens recovered from pirate ship *Whydah*	135
9.9	Possible gaming token, Valley Forge	136
9.10	Possible gaming tokens, Monmouth Battlefield State Park	136
9.11	Musket balls with worm marks, possibly gaming tokens, Valley Forge	137
9.12	Musket ball with flat spot found next to gaming die, Valley Forge	137
9.13	Partially flattened musket ball from American camp site, New Jersey	138

9.14	Flattened musket ball, Valley Forge	138
9.15	Shaped musket ball, Valley Forge	138
9.16	Lead pencil, Monmouth Battlefield State Park	139
9.17	Lead pencil, Newark, Delaware	139
9.18	Whizzers excavated in New York City and Youngstown, New York	139
9.19	Lead net sinkers	140
9.20	Musket balls converted into lead sinkers by pirates	140
9.21	Musket ball with hole for fishing line	141
9.22	Lead sinker, Valley Forge	141
9.23	Partially melted musket ball, Valley Forge	142
9.24	Partially melted musket ball, Continental army camp site, New Jersey	142
9.25	Two lead "wheels," Valley Forge	143
9.26	Musket ball with carved "face," Valley Forge	143
9.27	"Modern" bullet that resembles an impacted musket ball	143
9.28	Modern bullet, Valley Forge	144
10.1	Bird shot, buckshot, and musket balls	145
10.2	Bar of lead buckshot made in a gang mold	146
10.3	Eighteenth-century mold-cast buckshot	147
10.4	Bird shot with concave dimples from wrecked sailing vessel	147
10.5	Experimental results of dropping molten lead alloy into cold water	148
10.6	Experimental shot and actual shot from shipwreck	148
10.7	Teardrop-shaped experimental shot and actual shot from shipwreck	148
10.8	Reproduction colander for manufacturing small lead shot using Rupert method	149
10.9	Rupert shot recovered from *Queen Anne's Revenge*	149
10.10	Merchants' Shot Works shot tower and shot bag	151
10.11	Buckshot made with lead/pewter alloy	152
B.1	Modern musket balls with human chewing marks	161
B.2	Shot recovered from Pope's Fort	163
B.3	Artifact 93L13DS-4 with swine teeth impressions	164
B.4	Artifact 224-4-809 with shear mark probably caused by swine	164
B.5	Artifact 227-2-778 with swine molar impressions	164

B.6 Artifact 227-3-83 with swine bite marks — 165
B.7 Artifact 90M16RP4 with rodent gnaw marks — 165
B.8 Musket ball with major rodent chewing — 166
B.9 Side view of musket ball showing rodent teeth marks — 166
B.10 Artifacts with likely human bite marks — 167
B.11 Seventeenth-century swine mandible fragment — 167

D.1 BRAVO at Washington Memorial Chapel site, Valley Forge, Pennsylvania — 174
D.2 BRAVO at Blue Licks Battlefield State Park and Recreation Area, Kentucky — 174
D.3 BRAVO at Cooch's Bridge, Newark, Delaware — 174
D.4 BRAVO at Monmouth Battlefield State Park, New Jersey — 175
D.5 Members of BRAVO with Douglas Scott, Garry Wheeler Stone, and the author — 175

Maps

3.1 Interpretation of battle at Parsonage Farm orchard, Monmouth Battlefield State Park — 65
5.1 Cylindrical shot, Sutfin orchard, Monmouth Battlefield State Park — 80
6.1 Canister shot and grapeshot, Monmouth Battlefield State Park — 95
6.2 Iron case shot and lead canister shot, Monmouth Battlefield State Park — 97

Tables

2.1	Calculated density of American Revolutionary War musket ball lead	26
2.2	Typical eighteenth-century firearm bore sizes (or caliber)	31
5.1	Cylindrical shot excavated from the *Whydah*, 1717	81
7.1	Major elemental analysis of three musket balls from Monmouth Battlefield determined by X-ray fluorescence (XRF)	111
8.1	Calculated density of American Revolutionary War musket ball lead/tin alloy	117
8.2	Major elemental analysis of three "pewter" musket balls determined by X-ray fluorescence (XRF)	119
B.1	Monmouth Battlefield shot with unidentified teeth-related distortion	164
B.2	Specimens from Monmouth Battlefield likely mauled by swine leaving distinct teeth marks	165
B.3	Specimens from Monmouth Battlefield with likely human teeth marks	166
C.1	Small shot sizes and descriptions	170

Foreword

DAVID GERALD ORR

FOR OVER TWENTY YEARS I have known the work of Dan Sivilich. I first heard him present a paper summarizing his research at Monmouth Battlefield in New Jersey with Garry Wheeler Stone at a conference in Philadelphia, and I wanted to know more. Since the pioneering work of Doug Scott at Little Big Horn Battlefield, I had been aware of the opportunities presented by controlled metal detecting in the interpretation of what we know now as "fields of conflict." I was eager to use this technique as part of my own work at Valley Forge. We planned our first corroborative program there when I left the National Park Service in 2006 and began my full-time position at Temple University. Dan worked with a volunteer group known as BRAVO (Battlefield Restoration and Archaeological Volunteer Organization), made up of dedicated individuals who gave their time freely to assist parks and associated cultural resource managers in archaeological research. Much of the material in this book has its origin in those surveys.

In a larger sense (that is, beyond my long friendship with Dan), his work has illustrated several significant aspects of historical archaeological research. The first is the establishment of corporate data goals: in this case the study of the ubiquitous musket ball found at most revolutionary sites. Methodically and systematically, Dan used the musket ball as an eloquent spokesman for the interpretation of revolutionary battles and camps. It was the perfect example of Jim Deetz's "small thing forgotten." In this work Dan has let the musket ball act as an important and vital primary source for the understanding of eighteenth-century warfare. Second, Dan has given an important public archaeological dimension to the excavation of revolutionary sites. Interested and dedicated BRAVO volunteers are infectious in their enthusiasm for archaeology! They provide us with a host of linkages to an engaged public, which is necessary for effective and sustained preservation programs.

Our profession needs more in-depth looks at such artifact genres, be they buttons, needles, pipe stems, ceramic types, or tin cans. The time invested in such activity will restore much of the world we have previously lost.

Preface

When I announced to my wife that I was going to write a book about musket balls, she responded that it would not take long: "one page and one picture"! As I began writing, we were both surprised to see how many photographs I already had of many different variations. In 1987 I first started working on American Revolutionary War battlefield sites, and I could find little work done on studies of musket balls. There was no Internet, so data was limited to physical library searches. Over the past several decades, I have spent much time researching and identifying eighteenth-century and earlier lead projectiles and analyzing their characteristics, sizes, and other uses, such as being transformed into dice and other gaming pieces. This book is a compilation of those studies.

A great deal of information about a site can be learned by correctly identifying musket balls. The purpose of this book is to introduce the reader to the many types and shapes that musket balls come in. There are probably thousands of other types of musket ball forms still to be considered, so this book is meant to be a guide and is still a work in progress. Interpretations were made based on available information. Future analytical techniques may provide more accurate data, which may change some of my hypotheses.

The majority of the artifacts discussed in this book have been excavated from numerous military sites, but most were found at Monmouth Battlefield State Park and surrounding areas in Freehold and Manalapan Townships, New Jersey (the location of one of the largest and longest battles of the American Revolution, which occurred on June 28, 1778). As a result, this book deals primarily with American artifacts from the eighteenth century but touches on earlier European ordnance for general interest.

Battlefield archaeology is a relatively new science that has come into being because of the metal detector. Battles usually took place over very large areas that cannot be efficiently excavated by classical archaeological techniques. In August 1983, an accidental grass fire cleared much of the battle area at Little Bighorn Battlefield National Monument. Little time was available to conduct much archaeology, so Douglas Scott, PhD, made the critical decision to use volunteers with metal detectors to locate artifacts from that iconic battlefield; by doing so, he redefined the last moments of General George Armstrong Custer and the Seventh Cavalry (Scott 2013; Scott, Fox, and Connor 1989; Scott, Fox, and Harmon 1987), and the concept of modern battlefield archaeology was born. This concept has been adopted worldwide. For example, Glenn Foard, PhD, of the University of Huddersfield, United Kingdom, has done a great deal of work using metal detectors at English Civil War sites (Foard 2005, 2009).

All of these early archaeological surveys had one common goal: to define and interpret an area of military activity. The most common artifact found was the projectile. Scott had the good fortune of working on a site where modern bullets had been fired from rifled weapons. This allowed for an excellent forensic analysis of rifle land and groove signatures on bullets and firing pin signatures on brass cartridges. But what about earlier wars when simple musket balls were fired from weapons that left little or no identifiable markings? I am an engineer by vocation and have always looked to solve problems with mathematics and probabilities. When I dug my first musket ball, the immediate question that came to mind was: who fired it and at what target?

Unless otherwise noted, all photographs and graphics have been generated by (and are copyrighted by) the author. All diameters shown in the photographs are the measured or calculated diameters of the original spherical musket balls before alterations.

Note that it is illegal to remove any artifacts from all federal, state, and local protected historic sites. All of the work done by BRAVO has been with authorized archaeological permits and permissions. All artifacts illustrated in this publication have been recovered either by professional archaeologists or by avocational archaeologists on private property with permission from the owners or on public lands with no restrictions about metal detecting. The locations of all artifacts in this later group have been properly recorded. I have personally examined and documented the artifacts shown in this book that were found by the avocationalists. Many of these artifacts have been donated to local historical societies. The musket balls from the Seneca Iroquois village (figure 2.37) were properly documented by the New York State Archaeological Association, the records of which are on file at the Rochester Museum and Science Center, Rochester, New York.

Acknowledgments

ONE OF THE MOST SIGNIFICANT INFLUENCES on my life came from my best friend Ralph Phillips, who taught me nearly everything I know about archaeology, changing my life of relic hunting to conflict archaeology. Starting in 1987, many of the early excavations at the 1778 American Revolutionary War Battle of Monmouth were conducted by Ralph and me, digging in all types of weather. It was then that we met Nancy MacNeill and Bill Rainaud, two New Jersey State Park rangers, who encouraged us to work and to begin a volunteer group and who introduced us to the respected archaeologist Garry Wheeler Stone, PhD, historian at Monmouth Battlefield State Park. Much of our work would not have been possible without his guidance, trust, and forward thinking. In a time when many archaeologists were adamantly opposed to the use of metal detectors, Garry saw the benefit of using them as tools to excavate large battlefield areas and took a chance on letting us work at Monmouth Battlefield starting in 1990. The results have been spectacular.

However, little of this would have been possible without the thousands of hours volunteered by the members of the Battlefield Restoration and Archaeological Volunteer Organization (BRAVO) and its predecessor groups, the archaeology committees of the Friends of Monmouth Battlefield and the Deep Search Metal Detecting Club. My colleagues combed fields, bagging and tagging artifacts, point-proveniencing their locations, and cleaning and numbering the finds. I am very grateful for their time, dedication to archaeological knowledge, and friendship. Neal Barton, professional land surveyor and surveyor general of BRAVO, taught me everything I know about surveying that I use to map artifact locations.

In the 1980s, while we were working one end of the farm (which was then private property) where we discovered a major section of the Battle of Monmouth, three other gentlemen were relic collecting at the other end. Many years later our paths crossed again, and they joined BRAVO. They unselfishly donated their entire collection of artifacts from that farm to the state park, which significantly contributed to the information about the battle. Some of the

ACKNOWLEDGMENTS

photographs in this book are from that collection. A very special thank you goes to Dick Harris, the late Ted Harris, and Carlo Iovino.

In 1995 I attended my first Society for Historical Archaeology annual meeting, in Washington, D.C. At that conference I met many archaeologists who helped me expand my knowledge in the field of battlefield archaeology. Douglas Scott (PhD) was the first to use metal detectors and systematically map a battlefield. He is known worldwide for his groundbreaking studies at the Little Bighorn Battlefield National Monument. He truly is the father of battlefield archaeology.

A great deal of credit and thanks is due to Lawrence Babits (PhD). He proofread the original manuscript, correcting much of the grammar and insisting on many citations, which made this book an outstanding treatise on musket ball analysis.

Dana Linck contributed original work on the analysis of deer-chewed bullets. He and I have worked on numerous projects together but specialized in the New York Hudson Highlands (West Point, Peekskill, and Fort Montgomery).

A great deal of support came from many archaeologists I met at numerous conferences. Jo Balicki, Wade Catts, the late John Cavallo (PhD), Charles Haecker, Richard Hunter (PhD), Silas Hurry, Steven Potter (PhD), Michael Pratt (PhD), Gerard Scharfenberger (PhD), David Starbuck (PhD), Bruce Sterling, Elise Manning-Sterling, Tim Riordon (PhD), Richard Veit (PhD), Rebecca Yamin (PhD), and many more have been supportive of my work and have helped me transition into the world of professional archaeology by offering advice, encouragement, and constructive criticism.

Henry Miller (PhD) of Historic Saint Mary's City, Maryland, provided a great deal of his expert knowledge on the topic of animal- and human-chewed musket balls. His paper on this analysis is included as appendix B.

At one conference in Toronto, Adrian Mandzy (PhD) from Morehead State University, Kentucky, approached me and asked if I would be willing to help train graduate students how to use metal detectors on a 1649 battlefield. I remarked that I knew of no 1649 battlefields in the United States. He responded that I was correct—this was the 1649 Battle of Zboriv in Ukraine. Being of Polish and Ukrainian descent, I jumped at the opportunity and got firsthand experience excavating a European battlefield. Adrian and I became good friends and have worked together on several American battlefield projects. Much of the organization of this book was done with his guidance.

Many altered musket balls were found at a newly discovered camp site at the 1777–78 Valley Forge encampment. This effort is part of a graduate program project run by Temple University at the Washington Memorial Chapel, in which BRAVO assisted in locating key areas to excavate further. This project was made possible by David Orr, PhD (and doctoral advisor at Temple University), and his students Carin Boone and Jesse West-Rosenthal. Much gratitude goes to the Washington Memorial Chapel and Heritage for allowing the study to be conducted.

I cannot begin to thank the many people who contributed photographs and data, such as Bly Straube of the Jamestown Rediscovery Project in Virginia and Silas Hurry of Historic Saint Mary's City in Maryland.

Bill Ahearn, a good friend and fellow author, supplied a wealth of information and excellent photographs from his book *Muskets of the Revolution and the French and Indian Wars* that helped make this project happen.

Most of the European artifact photographs and descriptions were provided by my many friends in CAIRN (Conflict Archaeology International Research Network). Glenn Foard, Tony Pollard, and Tim Sutherland of the United Kingdom have provided much help and friendship over the years with artifact photographs and comparative data that have been invaluable. Glenn Foard has especially provided a great deal of knowledge of seventeenth-century ordnance based on his work published as *Battlefield Archaeology of the English Civil War*. I cannot say enough about the cooperation from my many European friends, such as Jakub Wrzosek of the National Heritage Board of Poland; Xavier Rubio-Campillo, postdoctoral researcher at the Barcelona Supercomputing Centre, Computer Applications in Science and Engineering; and Bo Knarrström of Sweden.

I owe a great deal of gratitude to many cultural resource management companies that have contributed to BRAVO over the years, especially John Milner Associates of West Chester, Pennsylvania; Hunter Research, Inc., of Trenton, New Jersey; and Richard Grubb and Associates, Inc., of Cranbury, New Jersey.

Much of my working knowledge of soldiering in the Revolutionary War has come from being involved in reenacting as a member of Mott's Artillery. John Mills, historian at Princeton Battlefield State Park, New Jersey, has taught me much about eighteenth-century field artillery. A great deal of information on how weapons were loaded and fired came from Jim Stinson, my friend and fellow Rev War reenactor with Proctor's Artillery.

Much credit must go to the State of New Jersey, Department of Environmental Protection, Division of Parks and Forestry, to the New Jersey State Historic Preservation Office, and to the New Jersey State Museum for years of approval and support. The Monmouth County office of GIS (Global Information Systems) donated all the digital aerial photographs to BRAVO for mapping the battle. The Monmouth County Historical Association provided much historical data about the Revolutionary War.

Many thanks go to the New York Office of Parks, Recreation and Historic Preservation and to the Fort Montgomery State Historic Site, particularly to Fort Montgomery Historic Site Manager Grant Miller, to State Historic Preservation Office (SHPO) archaeologists Paul Huey, Richard Goring, and Joseph McEvoy, and to Col. James Johnson (West Point historian, retired) for their

ACKNOWLEDGMENTS

participation in the Fort Montgomery project. Douglas R. Cubbison, Matthew Fletcher, and Paul Ackermann of the United States Military Academy (USMA), West Point, New York, arranged for and assisted in several archaeological surveys at the academy, which also yielded a significant number of artifacts.

Many farmers who allowed us access to their plowed fields—especially Charles, Lydia, and Jim Wikoff, tenant farmers at Monmouth Battlefield State Park, and Scott Applegate of Battleview Orchards in Freehold, New Jersey—often allowed us to tiptoe through their freshly planted fields to get in one more day of discovery.

One of the most unusual surveys that BRAVO conducted was at the Revolutionary War Battle of Bluelicks in Kentucky. It was the brainchild of Adrian Mandzy and required many areas of approval. My thanks go to the Kentucky Department of Parks, especially John Downs, and to the Kentucky State Nature Preserves Commission for a great deal of effort. It happened that the area to be surveyed was in a restricted area because of having one of the world's most rare plants—Short's goldenrod. Our every step literally had to be monitored by Joyce Bender and Zeb Weese of the Kentucky State Nature Preserves Commission to ensure that we did not damage any plants. They became interested in the project and worked hard to allow us to retrieve artifacts from the battlefield. Stephen McBride of McBride Preservation Service participated in the metal-detecting survey and served as a consultant for the project.

The late Ken Kinkor, former historian and museum curator of the Whydah Pirate Museum in Provincetown, Massachusetts, shared much of his information and theories on shot used by pirates and allowed me access to artifacts from the Whydah project. This provided valuable data for analyzing cylindrical musket balls. He spent a great deal of time debunking some of the myths about pirates and provided an accurate historical record.

I gave a paper at a conference on the lead shot that was recovered from a shipwreck discovered in New York City during the reconstruction of the World Trade Center. I was contracted to identify the shot and other artifacts in an effort to help date the ship. After the paper, I sat next to Paul Huey, a friend and archaeologist retired from the New York State Office of Parks. He whispered to me, "Look at Rupert shot." His suggestion completed my analysis and taught me something new.

Thank you to archaeologist Steve Warfel and to Janet Johnson (the State Museum of Pennsylvania) for their assistance in providing information and photos of items from French and Indian War sites in Pennsylvania.

Many thanks go to Joe Gagliardino for helping to count, weigh, and classify the large quantity of bird shot found on the World Trade Center shipwreck.

A great deal of gratitude and respect goes to the staff of the many historical sites mentioned in this book. They provided photographs, shared data, and allowed me to use their resources for this publication.

I am indebted to and appreciative of the anonymous peer reviewers of this publication for spending hours in reading, correcting, and improving it from a rough draft to a refined work.

There have been so many people involved in this project, it is difficult to thank them all individually; I apologize if I missed anyone.

The greatest thanks must go to my family. My fantastic wife, Lea, never complained about me wandering off to digs and meetings and spending a small fortune on my hobby. My daughter, Michelle, began digging with me when she was nine. She now has a BA in anthropology and in history and an MS in life science and has received her PhD for her thesis on American fortifications during the Second Seminole Indian Wars. My son, Eric, has been in the field and is a fellow Rev War reenactor. He has helped most recently with data analysis of locating artillery positions based on projectile scatter. He has a BS in Surveying Engineering Technology and does nearly all of the current archaeological surveying and mapping artifact locations for BRAVO.

Introduction

Douglas D. Scott

Most North American archaeologists who do fieldwork have found a lead bullet at one time. Bullets are ubiquitous in the American landscape for the simple reason that people shoot guns and have done so since Europeans stepped on the New World's shores in 1492. Firearms were used not only for offense and defense but also, more regularly, for the taking of meat for sustenance. Guns are still widely used today for hunting and leisure activities, such as plinking and formal target shooting, and, of course, in war.

Archaeologists finding bullets, whether spherical balls or conical bullets, ordinarily record them and describe them in their reports. Those descriptions often are simply a length, diameter, and weight with perhaps some speculation on the bullet's origin or use. Archaeologists more knowledgeable about firearms usually cite some other report in which bullets are described or one of the many collector books on firearms, cartridges, and bullets to make a more informed decision.

Conical bullets are relatively easy to identify and describe. There are many fine references on conical bullets, which became popular around the time of the American Civil War. Self-contained cartridges also appeared at that time and largely supplanted the spherical or round ball. References, likewise, abound on the history of cartridges, and dates of introduction and use are relatively easy to determine for almost all conical bullets and cartridges.

It is rare, however, for archaeologists or amateur collectors who report their findings to do much more than describe the lead spherical or round balls they find on their sites. Other than measuring diameter and weight (which generally equate to the caliber of the gun the balls were fired from or meant to be fired from), few archaeologists go beyond the basics with spherical balls. Daniel Sivilich's work on how to analyze a simple lead sphere, expounded in this volume, will change how we view musket balls as an archaeological artifact type.

INTRODUCTION

I had the privilege of meeting Danny at the 1995 Society for Historical Archaeology's annual meeting and conference. His interests in conflict and battlefield archaeology were what brought us together. We enjoyed then and in the ensuing years many conversations and exchanges of ideas such as the role of firearms in conflict and what happened to bullets as they terminated their flight trajectory, either from striking a target or when they reached their terminal velocity. Most of our intense conversations were held in conference hotel bars, the venue where great snippets of wisdom are routinely exchanged during scholarly meetings.

Over the years many people who shared similar interests joined in our conversations, leading to yet others who have contacted Dan. Some of those contacts resulted in new data for Dan to assimilate, much of which is reflected in this book.

Musket Ball and Small Shot Identification is a milestone in the study of firearms projectiles. Dan Sivilich neatly explores the history and development of firearms technology and the concomitant development and use of spherical projectiles. He clearly explains how musket balls were made—and most were not simply cast in a mold over some backwoods campfire. He touches on the development of ammunition manufacturers as he explores and describes how musket balls were made over time.

Musket balls had an intended use as projectiles, or did they? The details Dan is able to draw from a musket ball—how it was made, how it was loaded, what happened to it when it was fired, what did it hit—will make anyone reading this book rethink what can be learned from looking at a simple sphere of lead. Dan has amply illustrated the volume to give the reader a very visual as well as textual view of what can be learned from studying a musket ball, beyond determining its caliber. Every mark on a musket ball is a telltale trait that can be interpreted to aid in understanding what happened to that particular bullet in its short use-life sequence.

The chapter titled "What Did It Hit?" is a real eye-opener. In illustrating and describing real artifacts that hit something, Dan takes a step most researchers do not, and that is conducting experimental research employing live fire at similar targets. Normally this is the purview of law enforcement–based firearms examiners—testing external ballistics of firearms and their projectiles. But they seldom deal with antique weapons systems, so much of their literature is not relevant to solving specific questions relevant to the smoothbore musket era. Dan's results displayed with a "real" artifact clearly demonstrate the value of experimental archaeology in advancing the interpretive potential of musket balls. This is especially true in his chapter "Chewed Musket Balls," which helps demystify the origin of most chewed musket balls found on archaeological sites.

Likewise, the chapter on altered musket balls shows that those who war will find a means to improve the lethality of their shot. This work demonstrates that field experimentation and alteration of ammunition are nothing new. Dan's historical research also shows that there were concerns with the civility and legality of using altered musket balls in the past, and that is nothing new in warfare either. His work also demonstrates that there are rules that a society at a given time is expected to observe even in war, and persons who deviate from those rules are often called to task.

Bullets may be projectiles, but they are also the raw material of gaming pieces, ersatz pencils, toys, and just something to whittle on. Soldiers put lead bullets to many uses in camp. As a readily available and readily worked raw material, it kept the hands of the idle soldier busy. These modified bullets tell the story of boredom in camp and the creative nature of man to make bullets into a metaphorical plowshare, to borrow an expression.

Musket Ball and Small Shot Identification is an important contribution to the field of firearms artifact studies. Dan Sivilich takes what was a simple artifact that was easily and succinctly described to new depths of analytical power. It will no longer suffice to simply describe a musket ball as a spherical lead object of a certain diameter and weight. There is much more to learn about its past function by reading the marks of manufacturing, manner of loading, and dings and nicks of impact that Dan Sivilich so clearly explicates in this volume.

Dan describes himself as a trained engineer who is an avocational archaeologist. This volume demonstrates that he is also a consummate researcher and professional, and this work changes how we see and will analyze spherical lead projectiles in the future.

Musket Ball and Small Shot Identification

CHAPTER 1

THE GUN AND EARLY PROJECTILES

ON OCTOBER 12, 1492, three small ships arrived at an island in the Bahamas, and the captain named the island San Salvador. Christopher Columbus and his men waded ashore and changed the course of weaponry in North America. They were carrying matchlock muskets that fired lead musket balls.

Of course, the history of projectile weaponry goes back much farther than that. When the first hominid picked up an object such as a stick or a rock and threw it, the concept of a projectile was born. The projectile could be used for

1.1. Landing of Columbus, oil on canvas, 12'×18', by John Vanderlyn. Vanderlyn was commissioned to create the painting in June 1836; it was installed in the U.S. Capitol rotunda in 1847. *(Image courtesy of Architect of the Capitol website, www.aoc.gov.)*

MUSKET BALL
AND SMALL SHOT
IDENTIFICATION

either defensive or offensive purposes. Pointed sticks became spears, and rocks became blunt-trauma ballistic projectiles. Long before Sir Isaac Newton developed the equation $F=ma$ (force is equal to mass times acceleration), early man learned to increase the force of his projectile by external means, such as artificially extending the length of his arm. He came up with the atlatl, or throwing stick, which increased the length of his throwing arm and thus also increased the centripetal force and thus the velocity for greater distance and penetration. The invention of the bow and arrow became the next major development. Bone arrowheads dating to 61,000 years ago have been found in a cave in South Africa (*Britannica* 2014).

Another ancient projector is the sling. This is a leather or fabric "cup" or cradle with long lengths of cord on either side. A projectile was placed in the cradle, the two lengths of cord were brought together, and the assembly was twirled to increase centripetal force. When one of the cords was let go, the projectile was released at a lethal speed. At some point, man learned that increasing the mass (or actually the density) of the rock caused it to hit with greater force. Extracted lead from the mineral galena was cast or hammered into shapes. Being much denser than most rocks, lead made an excellent projectile material for use in slings. Eventually, the shot was made into an ellipsoid, or almond, shape to fit the sling pouch. Numerous examples of this kind of shot have been excavated at Roman, Greek, and Middle Eastern military sites. Several examples are shown in figure 1.2.

The bow was the weapon of choice for long distances. The use of metal arrowheads increased penetration, but hitting a target required a skilled and strong archer. Longbows in the Royal Armouries in Leeds, England, recovered from the shipwreck *Mary Rose* (1545) required more than 150 pounds of force to draw the string back, according to the museum display. The crossbow increased penetration power and required very little training, as it was a point-and-shoot

1.2. Left: Roman lead sling shot from the AD 9 Battle of Teutoburg Forest, at Kalkreise Hill, Varusschlacht, Germany *(photograph by Michelle Sivilich, PhD). Right:* The left two specimens are Greek, and the right two are Roman lead sling shots *(from the collection of James Legg, South Carolina Institute of Archaeology and Anthropology; photograph by James Legg).*

weapon. It had a stock and trigger, the design of which was later used for some of the earliest shoulder arms.

Mankind continued to find ways to kill or hit a target at greater and greater distances. The crowning achievement still in use today is the gun. Appendix A lists some of the significant dates in the development of guns that led to modern firearms. But what were the projectiles fired from these early weapons? Some of the earliest known musket balls have been excavated in the Czech Republic. Seven iron bullets coated in lead were excavated at the castle of Scion, which was destroyed in 1437 at the end of the Hussite war. Cylindrical bullets were also found at the 1420–21 military camp in Kunratice (Janská 1963: 244).

A lead-covered stone musket ball (figure 1.3) was excavated at the site of the 1649 Battle of Zboriv, Ukraine, near the location of a Cossack camp. It had an overall diameter of 0.65 inches. Similar projectiles are reported to have been found near the site of the 1648 Battle of Pyliavtsi, Ukraine (Adrian Mandzy, personal communication, October 9, 2012).

Early musket balls were made from a variety of materials. Glenn Foard, who holds a doctorate in archaeology and supervises doctoral students at the University of Huddersfield in England, reported seeing glass musket balls and artillery projectiles in a fifteenth- to sixteenth-century exhibit at a castle in Styria, Austria (Foard, personal communication, August 3, 2012). Figure 1.4 depicts a glass musket ball excavated near the fourteenth-century Castle Puck in the province of Pomorze Gdańskie (Pomerelia), Poland (Mandzy, personal communication, November 2, 2012). An archaeological survey was conducted at the site by the Institute of Archaeology in cooperation with the Institute of Archaeology and Ethnology of the Polish Academy of Sciences in Warsaw. Iron points from crossbow bolts, arrowheads, stone cannonballs, and lead musket balls were also excavated.

Over the centuries, lead became the material of choice for projectiles. Because lead has a low melting point (621.5°F, or 327.5°C), it can be melted over a campfire and made into a musket ball using a simple mold. The spherical lead musket ball became the preferred projectile until the mid-nineteenth century.

What were the typical weapons used to fire musket balls and other projectiles? There are many technologies associated with the evolution of early arms, and these are far beyond the scope of this publication. A cursory overview of the most common types of guns in use from the fifteenth to nineteenth centuries follows to provide necessary background.

THE GUN AND EARLY PROJECTILES

1.3. Seventeenth-century lead-covered stone musket ball from Zboriv, Ukraine. *(Photograph by Adrian Mandzy, PhD.)*

1.4. Seventeenth-century glass musket ball from Castle Puck, Poland. *(Photograph by Adrian Mandzy, PhD.)*

MUSKET BALL
AND SMALL SHOT
IDENTIFICATION

1.5. The earliest known picture of a gun. *(Drawing from Ulrich Bretscher's Black Powder Page, www.musketeer.ch/blackpowder/history.html, used with permission.)*

1.6. Two guns, or "handgonnes," found in the debris of the Freienstein castle, Canton Zurich, Switzerland. *Left:* Iron arquebus, cal. 29 mm (1.14"), with the remains of its hook, which served as a bench rest and an antirecoil device, and the enforcement ring of the wooden stock (now missing because of decay). *Right:* Iron handgonne, cal. 18 mm (0.71"). The breech with its touchhole is missing. It possibly exploded and only the cracked barrel part remained. The barrel cracked along its weld seam. *(Photograph by Martin Bachmann, Archaeological Service of the Canton, Zurich, Switzerland.)*

The earliest known representation of the use of a firearm comes from the manuscript *De nobilitatibus, sapientiis et prudentiis regum*, by Walter de Milemete. It shows a jug-shaped cannon firing an arrow (figure 1.5).

By the late fourteenth to early fifteenth century, the concept of a handheld gun—or "handgonne," as it was referred to—was developed. Figure 1.6 portrays two early shoulder arms recovered as follows:

> In the course of an archaeological excavation of the castle of Freienstein, situated in the namesake village of the Canton Zürich, the remains of two handgonnes were found in 1975. According to historical records, the castle burned between 1429 and 1474 and wasn't reconstructed after this. The gonnes were found in the debris together with some swords, locks, and kitchen utensils. Hence the gonnes are certainly older than 1474, probably ca. 1380–1400.
>
> Though, according to the oral history of the locals, the castle became a victim of the "Old Zürich war" (there was also a new one), it's likelier [that] the castle burned accidetally [sic]. This theses [sic] is underlined by the fact that expensive artefacts like swords, armours plus a pan from brass were found among the debris, too. Usually castles were looted before setting them afire. (Bretscher 2009)

The hand cannon was cumbersome and not very accurate. Combining the concept of a gun barrel with the stock and trigger mechanism of a crossbow led to the development of the matchlock in 1425. The matchlock had a simple firing mechanism (figure 1.7) known as a lock (Koenig 2012). Slow match, a rope impregnated with saltpeter (potassium nitrate) to make it burn slowly and evenly, was held by an S-shaped lever called a serpentine. The vent hole on top of the barrel was changed to a touchhole located on the side of the barrel near the breech to leave a clean line of sight along the barrel. A small pan fastened outside the hole was filled with black powder. The pan had a cover on a hinge to keep the powder from falling out while the musket was being carried and, more importantly, to keep stray sparks away from the priming powder. Firing a matchlock after it was loaded and primed required that the pan cover be pivoted out of the way, and as the trigger was squeezed, the burning slow match was lowered into the pan. If the vent was blocked, one would experience a *flash in the pan*—the origin of the idiom—and the gun would not fire. This gun was somewhat awkward, because it required a length of slow match that had to constantly be fed into the serpentine jaw as it burned shorter. At that point in time, a gun consisted of a *lock, stock,* and *barrel* (leading to another common expression).

Matchlocks were taken to the New World by early explorers during the colonization of the Americas. Parts from matchlock muskets have been excavated from the 1607 Jamestown fort in Virginia, the location of the first permanent British colony in the New World (Beverly Straube, personal communication, November 15, 2012).

The next major innovation in early arms was the development of the wheel-lock musket in 1509. It had a viselike clamp, called a dog's head, on the serpentine. This clamp held a piece of iron pyrite (also known as fool's gold) and, later, flint. A spring-loaded wheel with a coined edge was wound with a spanner (wrench) applied to an exposed square crankshaft. An internal ratchet mechanism was connected to the trigger and the wheel, which was kept in place with a taut main spring. Like the matchlock, the wheel lock had a pan with a cover, but on the wheel lock, the cover was cammed to the crankshaft. After the gun was loaded, the dog's head was lowered, with the pyrite resting on the pan cover, and was held there by a strong spring. When the trigger was pulled, the wheel rotated quickly, the crankshaft cam opened the pan cover, and the pyrite contacted the rotating wheel, which acted like a grindstone. A spark ignited the priming powder, which

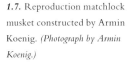

1.7. Reproduction matchlock musket constructed by Armin Koenig. *(Photograph by Armin Koenig.)*

then caused the gun to discharge. Although faster and more reliable than a matchlock, the wheel-lock mechanism was complicated and was usually made by clock makers. It was expensive, causing it to never achieve widespread military use. However, it was easy to use from horseback as a pistol or a carbine (Neumann and Kravic, 1989:5). A typical wheel-lock design is shown in figure 1.8.

The concept of striking steel with a piece of flint to make a spark was centuries old, but in the early sixteenth century this concept was incorporated in a simple gunlock called a snaphaunce (figure 1.9). The dog's head became known as a "cock" and held a sharpened piece of flint in a jaw. The flint was typically wrapped with leather or a small lead sheet to keep it from sliding in the jaw. The pan cover was connected to the locking sear and would open and close with the movement of the cock. A simple hardened steel striking surface, called a battery, replaced the complicated wheel. After the gun was loaded and the pan primed, the cock was pulled back into the firing position and the pan was closed. The battery was lowered on top of the pan lid,

1.8. Reproduction wheel-lock musket constructed by Armin Koenig. *(Photograph by Armin Koenig.)*

and the weapon was ready to fire. When the trigger was pulled, the pan lid automatically slid forward, exposing the priming powder. As the flint struck the battery, it generated a spark that fell into the priming powder in the pan. When the primer ignited, the flames traveled through the touchhole and into the gun's main powder charge at the breech of the barrel, causing the gun to fire.

MUSKET BALL
AND SMALL SHOT
IDENTIFICATION

1.9. Early seventeenth-century colonial snaphaunce lock. *(Photograph by Bill Ahearn.)*

1.10. A Model 1727 long land pattern British Brown Bess musket with banana-shaped lock. *(Photograph by Bill Ahearn.)*

By 1709, the battery of the snaphaunce had evolved into what was called a hammer, steel, or frizzen (the use of this latter term has been debated), which combined the battery and pan cover in one piece. This essentially became the flintlock musket (examples shown in figures 1.10 to 1.12). The cock had two positions: half cocked and fully cocked. The half-cocked position was a safety used to prime the pan and load the gun: the cock was in a locked position, and pulling the trigger did not cause it to fall. This is probably the origin of the expression "lock and load," because the cock had to be pulled all the way back to fire this gun. Thus, the admonishment "Do not go off half cocked" came into being. The flintlock was used in America beginning in the late seventeenth century and into the American Civil War era. This was the weapon of choice during the French and Indian War and the American Revolution. (One of the best books on this topic is Bill Ahearn's *Muskets of the Revolution and the French and Indian Wars* [2005].)

The British infantry used a smoothbore musket known as a Brown Bess during the French and Indian War, the Revolutionary War, and the War of 1812. During this time, the Bess evolved from a long land pattern to a short land pattern and an India pattern. Small changes were made in the lock mechanisms (see figures 1.10 and 1.11). Over time, the barrel lengths were shortened, but each model maintained a standardized 0.75" bore and took a 0.69" ball. They also used a standardized socket bayonet. One of the best reference books on this topic is Eric Goldstein and Stuart Mowbray's *The Brown Bess: An Identification Guide and Illustrated Study of Britain's Most Famous Musket* (2010).

The French were true innovators in early gun development. The French musket (figure 1.12), frequently referred to as a "Charleville," had the same lock design as the British. However, the British musket had the barrel and ramrod pipes pinned to the wooden stock, making it difficult to field repair. In 1763 the French developed metal barrel bands that integrated ramrod pipes and held the barrel to the stock. These barrel bands had a simple "push-button" spring that held them in place on the stock. When the spring buttons were pressed, the barrel bands could be slid off the stock and the barrel removed for cleaning or replacement. This made field repairs very easy. This concept was copied by Eli Whitney and all other major American gun manufacturers for musket and rifle design until well after the American Civil War.

THE GUN AND EARLY PROJECTILES

During the early years of the Revolutionary War, the Americans had very little in the way of standardized weapons. Many muskets and rifles were the guns that the farmer-soldiers brought from home. They lacked bayonets and did not use standard size musket balls, and the British knew this. The following was taken from an account of the Battle of Princeton, New Jersey, on January 3, 1777:

1.11. A Tower-marked Model 1777 military British Brown Bess musket. *(Photograph by Bill Ahearn.)*

It cannot escape the observation of any person who has attended to the circumstances of this war, that the number slain on the side of the Americans, has in general greatly exceeded that in the royal army. Though every defect in military skill, experience, judgment, conduct, and mechanical habit, will in some degree account for this circumstance, yet perhaps it may be more particularly attributed to the imperfect loading of their pieces in the hurry of action, than to any other cause; a defect, of all others, the most fatal; the most difficult to be remedied in a new army; and to which even veterans are not sufficiently attentive. To this may also be added the various make[s] of their small arms, which being procured, as chance or opportunity favoured them, from remote and different quarters, were equally different in size and bore, which rendered their being fitted with ball upon any general scale impracticable. (Dodsley 1778: 19)

1.12. A Model 1768 French military musket from the Charleville, France, arsenal. *(Photograph by Bill Ahearn.)*

Typically the early battles consisted of the British and the Americans firing a few volleys, after which the British would charge with fixed bayonets. Not having many bayonets, the American lines often broke and retreated. The lack of standardized weapons changed with the help of France. The French supplied the Continental army with standardized smoothbore muskets and bayonets, and this became one of the key factors leading to an American victory in the War for Independence. Being long-standing enemies of Great Britain, the French seemed eager to help the American cause:

> In 1776, the Massachusetts Board of War bought an unknown number of French muskets to arm the State troops. In total more than 20,000 foreign weapons, and possibly a good deal more, were imported by Massachusetts alone during the Revolution.
>
> As the conflict developed, the French and the Americans found common ground in their hatred for the British. In 1776, the Continental Congress sent Silas Deane and Benjamin Franklin to Paris. One of their duties was to arrange for the purchase of French arms. At first, clandestinely under the dummy trading company of Roderique Hortalez et Cie., and then overtly after France officially entered the war, the French supplied the needed munitions. In total, France supplied well over 100,000 firearms to the colonies and perhaps as many as 200,000. The price paid for these arms was less than half of that paid for the Committee of Safety contract muskets, and most of these muskets were obtained under a sort of Lend-Lease agreement that paid for the weapons with money borrowed in France. (Ahearn 2005: 175–76)

Each musket was marked with the arsenal name on the lock plate, but today's general description of a French musket used by the Americans is "Charleville."

> [T]he French Army regulation arm of the period, the Charleville model of 1763.... Together with other French regulation muskets made at the Royal Arsenals of Maubeuge, St.-Etienne, and Tulle, which differed only slightly in design, it was the finest military arm of its day. Manufactured with greater care and having an improved type of hammer and barrel securely fastened to the stock by bands instead of "pins" through lugs, it possessed greater durability, accuracy, and range than did the British musket, or the Colonial arms modeled from it, with which the Americans entered the war. The Charleville model was somewhat lighter than the British arm and its caliber was less, having a bore of about .69 inch. (Hopkins 1940)

By late 1777, these muskets had arrived at Valley Forge. General Friedrich Wilhelm von Steuben trained the troops to use the French musket and, more importantly, the bayonet. On June 28, 1778, both armies clashed at Monmouth

Courthouse (today, Freehold and Manalapan Townships), New Jersey. The American troops used their training and stood their ground when the British charged with fixed bayonets. The Continentals fixed their shiny new French bayonets and went hand to hand with the Crown forces. The Americans held the field at Monmouth, which was a political victory for General George Washington and a turning point of the war. General Henry Clinton, commander of the British army, moved the war to the south, which ultimately led to General Charles Cornwallis's surrender at Yorktown in 1781. The success of the American army was partly attributable to standardized arms and bayonets.

Since its formation in 2000, the Battlefield Restoration and Archaeological Volunteer Organization (BRAVO) has been working with numerous archaeologists in excavating early American conflict sites. One such site is the Washington Memorial Chapel in Valley Forge, Pennsylvania. Specifically, BRAVO has been working to identify features at a significant 1777–78 encampment area discovered by Temple University professor David Orr. Hut footprints and a camp kitchen have been recently excavated. BRAVO has been helping to locate activity areas by using metal detectors to recover artifact concentrations that can be further excavated for features and nonmetallic artifacts. On December 8, 2012, a metal-detecting survey was conducted by BRAVO with doctoral candidate Jesse West-Rosenthal and Dr. Orr supervising. Within minutes Jim Barnett of BRAVO recovered a very encrusted musket lock (figures 1.13 and 1.14) in a

THE GUN AND EARLY PROJECTILES

1.13. Musket lock at the excavation location at the encampment site at the Washington Memorial Chapel, Valley Forge, Pennsylvania.

1.14. Close-up of the excavated musket lock with heavy encrustation.

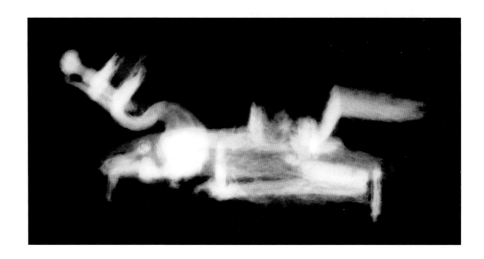

1.15. X-ray of the excavated musket lock showing the internal parts. *(Photograph courtesy of Maryland Archaeological Conservation Laboratory.)*

1.16. Both sides of the cleaned musket lock found at the Washington Memorial Chapel site, Valley Forge. *(Photographs courtesy of Maryland Archaeological Conservation Laboratory.)*

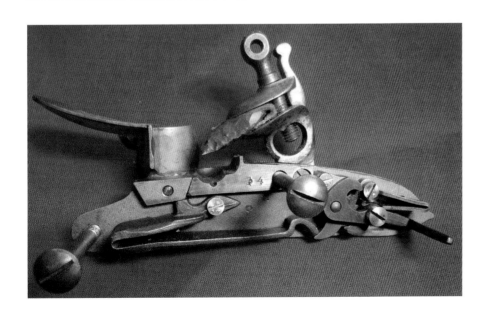

1.17. A reproduction French Charleville lock. This is very similar to the actual lock shown in figure 1.16. *(Photograph by Jim Stinson.)*

MUSKET BALL AND SMALL SHOT IDENTIFICATION

1.18. A Model 1728 French musket from the Ahearn collection. *(Photograph by Bill Ahearn.)*

potential trash midden. This artifact was taken to the Maryland Archaeological Conservation Laboratory for preservation, where it was also x-rayed (figure 1.15), cleaned using microabrasion (figure 1.16), and then conserved.

As a basis for comparison, Jim Stinson, fellow reenactor and eighteenth-century black powder weapons expert, provided a photograph of his reproduction 1777 French Charleville lock to show the internal mechanisms (figure 1.17). The brass plate to the left of the flint is a modern brass flash guard that would not have been on an original Charleville musket but is required for reenacting to prevent accidental powder burns from the pan and vent flash to a person standing next to you in a tight firing formation.

According to Bill Ahearn, an expert on Revolutionary War muskets, the lock excavated at the Washington Memorial Chapel site is possibly from a French Model 1728 musket but more likely from a French Model 1754 musket, which are very similar in general appearance; additional conservation may further clarify the identification. Notice the scribe line just to the left

1.19. Cache of decommissioned bayonets excavated at the Washington Memorial Chapel site, Valley Forge. *(Photograph by Anita Hermstedt.)*

of the cock in figure 1.18. The same type of marking is lightly visible on the excavated lock.

Other military artifacts have also been found at this site. On September 28, 2013, a single cache of thirty bayonets was discovered by Bill Hermstedt of BRAVO and excavated by Jesse West-Rosenthal (figures 1.19 and 1.20). Most of the thirty-two bayonets recovered at the site had been intentionally made unusable by crushing their sockets. This suggests that the old items may have been decommissioned once the soldiers received new French muskets and bayonets.

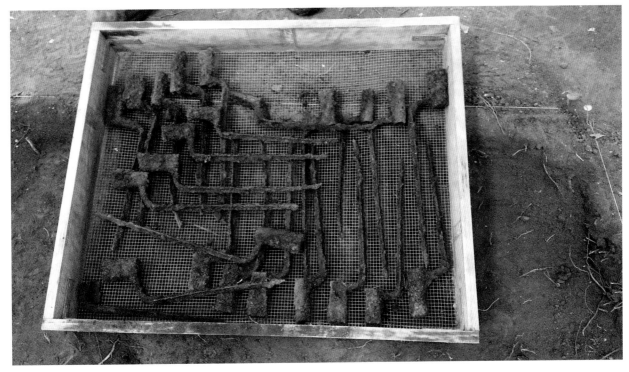

1.20. Twenty-four of the cache of thirty bayonets excavated at the Washington Memorial Chapel site, Valley Forge. *(Photograph by Glen Gunther.)*

Chapter 2

The Basic Musket Ball

WHAT IS A MUSKET BALL? A common description would be, "A small lead sphere designed to be fired from a musket, pistol, or rifle at a specific target with deadly force." Is this definition true? Are musket balls

- made of lead?   *Not always*
- spherical?   *Not always*
- used with a musket, pistol, or rifle?   *Not always*

So how does one identify artifacts found at early military sites that are or may have been musket balls? I will start with basic musket balls made of lead that did begin as spheres and were designed for use with a musket or rifle and will then examine variations.

Spherical Musket Balls

2.1. Artifacts found at the 1655–75 Dann site, a Seneca Iroquois village, now in the collections of the Rochester Museum and Science Center, Rochester, New York.

Musket balls are manufactured by pouring molten lead or another alloy into a two-part single- or multiple-cavity mold. After the lead cools, the mold is separated and the musket ball removed. The casting sprue is cut close to the ball, and any flashing around the mold seam is removed. During the early colonization and exploration years of the seventeenth century, lead was brought from Europe in ingots as shown in figure 2.1 (artifact 30). Pieces of an ingot were cut off and melted over a fire in an iron ladle (artifact 31). The molten lead was then poured into a mold (artifact 32). The artifacts shown in figure 2.1 were excavated at several seventeenth-century Seneca Iroquois village sites near Rochester, New York.

During the eighteenth century, musket balls and shot were both imported and locally made. Usually military musket balls would be put into paper cartridges with premeasured charges of black powder. Eighteenth-century molds were usually made of iron or brass, but crude molds made of soapstone and brownstone have also been found (Neumann and Kravic 1989: 190–93). Figure 2.2 shows a steatite (soapstone) mold currently at the Monmouth County Historical Association, Freehold, New Jersey. As can be seen, two different musket ball sizes and several sizes of small, or "buck," shot could be cast with this mold. Several buckshot cavities are ganged together to maximize the available capacity of the mold. The different musket ball sizes suggest that the owner had two weapons, possibly a Brown Bess smoothbore musket or large-bore pistol and a smaller-bore rifle or fowler. Note that the side of the mold is inscribed "17" on the right half and "76" on the left half.

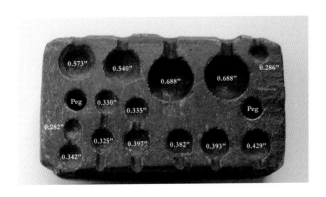

2.2. Steatite musket ball and small shot mold. *(Photographs by the author with permission of the Monmouth County Historical Society, Freehold, New Jersey.)*

The following terminology is used in describing weapons and musket balls:

Mold seam—A thin line around the circumference of all molded shot. Some molds were crude, and the two halves would not match exactly when closed. This resulted in musket balls that have two slightly offset halves.

Casting sprue—A small raised cylinder from the lead inlet channel in the mold. This is usually clipped off close to the surface of the musket ball, creating a small medial ridge.

Patina—Lead carbonate/oxide/sulfate. Shot that has been buried in the ground for some time develops a white lead oxide, lead carbonate, and lead sulfate coating (Rupert Harris Conservation 2013). Iron or other chemicals in the soil can change the color of the patina from white to tan to brown. Pine and oak trees produce high levels of tannic acid that can change the color of the patina to a dark reddish brown.

MUSKET BALL
AND SMALL SHOT
IDENTIFICATION

Diameter—The size of a musket ball (usually measured in inches). This is not the caliber of the gun.

Caliber—The diameter of the gun barrel bore.

Windage—The difference between the gun caliber and the ball diameter. Typically the windage is approximately 0.05–0.10 inches for military smoothbore muskets (Neumann 1967: 52).

If round musket balls are found in quantities in areas not known to have a conflict, this can indicate a camp site where possibly musket balls may have been cast or cartridges were made. If found in a battle area, they may indicate where soldiers were loading their firelocks or where soldiers fell and cartridges spilled from their cartridge boxes. Round musket balls found in a conflict area may also have been fired and hit a soft target, such as sandy soil, but careful examination of the surface characteristics will usually identify slight distortions in the lead.

Not all dropped musket balls have a visible mold seam or casting sprue. Unfired musket balls have been excavated at Revolutionary War British sites that do not have these two features. Most musket balls recovered from a British site in Middletown, New Jersey, had micro dimples on their surface similar to, but smaller than, that of a golf ball. Lead mining in the colonies was limited to a few small areas such as in Virginia (Burns 2005: 112). It is most probable that these musket balls were made in England, packed tightly in crates or barrels, and transported to the colonies by ship and then inland by wagons. The rough modes of transportation could cause the balls to bang together many times, causing the mold seams or casting sprues to be erased and micro dimples to develop (Sivilich 1996).

The example shown in figure 2.3 is from the collection of Monmouth Battlefield State Park, the site of the June 28, 1778, battle between the Continental army and the Crown forces. It is 0.667 inches in diameter, indicating that it may be British, but the intact mold seam and sprue cut suggest that it is of American origin. This musket ball was classified as probably not fired, based on the strong features and lack of deformation even to the thin mold seam line. Numerous examples of unfired musket balls have been found in non–conflict areas such as Valley Forge and other camp sites. An excellent example was found at the 1779 Battle of Briar Creek in Georgia. The Americans were camped near the creek and were preparing to meet the British when they were surrounded by the British and attacked. The large number of unfinished and unfired musket balls and lead waste excavated in one area suggest that the Americans were manufacturing lead shot (Dan Battle, personal communication, August 29, 2014), as shown in figure 2.4.

Unfired musket balls are fairly common at battlefield sites as well. Why would unfired musket balls be found on a battlefield? One reason is that the musket ball may have simply rolled out of the barrel. If the barrel was clean, the windage between the bore and the ball would be at its maximum. If the ball

THE BASIC
MUSKET BALL

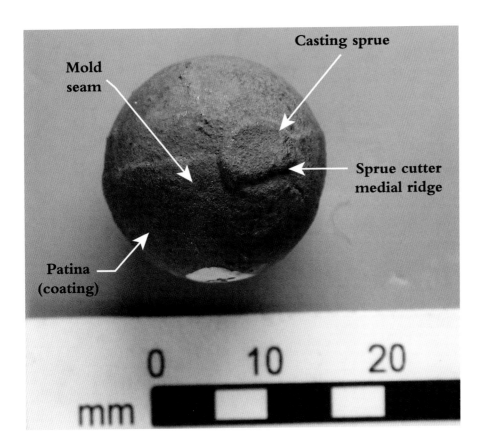

2.3. Musket ball common terminology showing mold seam and casting sprue, etc.

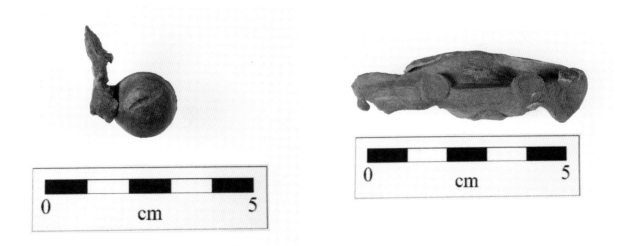

2.4. *Left:* Musket ball with unclipped sprue. *Right:* Lead waste from a two-musket-ball gang mold. *(Photographs by Daniel Battle.)*

was not wadded and the musket was tilted down, as would be the case in shooting downhill, the musket ball would have simply rolled out onto the ground. Another probable explanation is that they were dropped. Soldiers did not load their muskets using a powder horn to charge the musket and then drop in a loose ball as Hollywood would have us believe: this procedure would have been too slow, and variable powder charges would have created inaccuracies in hitting a target from shot to shot. Instead, soldiers carried premade paper cartridges that had an accurately measured powder charge and a musket ball (as shown by the reproduction cartridge in figure 2.5). This would guarantee that the musket ball would be propelled by the same force. Some cartridges may have also contained buckshot (as discussed later, with figure 2.16). The cartridges were carried in a cartridge box. In the heat of battle, it was possible for a soldier to accidentally grab two cartridges by their paper pigtails (figure 2.5). Rather than taking the time to try to put one back into the box, the soldier could have simply dropped the extra cartridge. Areas of firing lines have been identified at Monmouth Battlefield State Park because of large, linear concentrations of dropped musket balls. As stated earlier, cartridges could also spill out of a cartridge box if a soldier fell. Another possibility is that the musket ball might just have been oversized and could not fit in the bore and was discarded.

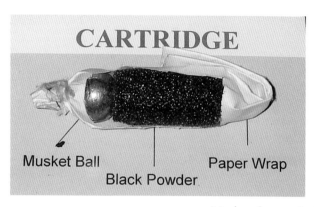

2.5. Reproduction military cartridge.

Musket shot analysis can significantly add to the historic interpretation of a site. Such an analysis was conducted by the author to help date the remains of a wooden ship found at the base of the World Trade Center during the reconstruction project (Sivilich 2011). A large number of shot was found among the collapsed timbers. Based on the types and sizes of shot, the author concluded that vessel was from the late eighteenth to the early nineteenth century.

One of the earliest archaeological sites in the United States in which large quantities of lead shot were excavated was in Jamestown, Virginia. This was the first permanent British colony, established in 1607. Beverly (Bly) Straube, the senior archaeological curator at the Preservation Virginia, Jamestown Rediscovery project, provided shot data from the first ten years of occupation of the Jamestown fort context along with the following information: "We have evidence for matchlocks, snaphaunces, and wheel locks with matchlocks being the most numerous. We have two lead gaming dice as well as one lead cube that has not been marked but appears to be a die in process. We often find the lead shot attached to the runner or with sprue attached. Shot cast in a mold that was misaligned has also been found" (Straube, personal communication, November 15, 2012).

A distribution diagram of 3,779 pieces of lead shot is shown in figure 2.6. The largest concentration is small lead scatter shot. This type of shot is typically named for the game animal that it is intended to kill, such as bird shot and buckshot. Unless musket balls were not dropped as much as the smaller shot, this chart indicates that the settlers were carrying shot for hunting. Because the fort was on a river, waterfowl were most likely abundant, and thus the bird-shot concentration is high. White-tailed deer, wild boar, and other mammals were also available for the food supply. This would explain the buckshot, which was usually used with a larger musket ball.

Unusual musket ball molding characteristics can help identify the general source of manufacture. In the case of the American Revolution, the presence of offset molded musket balls (like those shown in figure 2.7) usually indicates a mold originally of civilian origin.

Musket balls VF10-253 and VF10-255 (figure 2.7, *far left, left*) are from the Washington Memorial Chapel site, a 1777–78 Continental winter encampment site at Valley Forge, Pennsylvania. This site was occupied only by American soldiers. FtM10-947 (figure 2.7, *right*) is a musket ball that was excavated at Fort Montgomery, which was built by the American army on the Hudson River in New York and captured and burned by the British in 1777. VF10-253 is ovoid in shape with a strong mold seam. A more defined mold seam is an indication that the two halves of the mold did not seat with a good seal. VF10-255 has an offset mold seam, which means the mold did not line up properly. WF10S-002-28 (figure 2.7, *far right*) was recovered in the remains of a late eighteenth- or early nineteenth-century wooden sailing ship that was discovered while excavating

THE BASIC MUSKET BALL

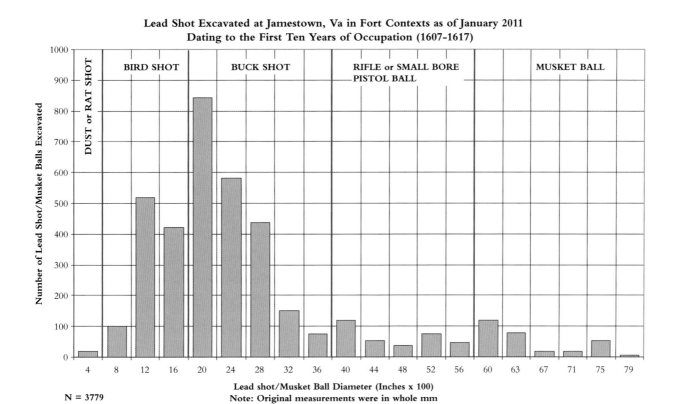

2.6. Diameter distribution of musket shot (*N*=3,779) from the first ten years of occupation at Jamestown, Virginia (1607–17), excavated in fort contexts as of January 2011. The original measurements were in millimeters and have been converted to inches (multiplied by 100).

MUSKET BALL AND SMALL SHOT IDENTIFICATION

2.7. Musket balls VF10-253 (*top left*) and VF10-255 (*top right*) excavated at the Washington Memorial Chapel site, Valley Forge. Musket ball FtM10-847 (*bottom left*) was found at Fort Montgomery, New York. Musket ball WF10S-002-28 (*bottom right*) was recovered from a late eighteenth- or early nineteenth-century ship discovered at the World Trade Center site in New York City.

the base of the destroyed World Trade Center towers (Sivilich 2011). It was cast in a mold that did not seat squarely, causing the musket ball to be made with two slightly offset halves. It is very similar to VF10-255. All of these musket balls are of different diameters and were recovered from three different sites, but all of them were made with poor mold alignment. This suggests that they were probably cast in crude molds like that shown in figure 2.2.

What type of musket (or rifle) did a musket ball come from? To determine the weapon it came from, one must know the diameter of the musket ball. If the ball is round, the diameter can be directly measured using a good set of calipers. When the diameter is known, the bore of the gun can be estimated by adding the estimated windage to the diameter. As mentioned earlier, the windage is approximately 0.05–0.10 inches for smoothbore muskets used in military service (Neumann 1967: 52). This is because of the fouling properties of black powder. Black powder is a mixture of charcoal, sulfur, and saltpeter. When a musket is

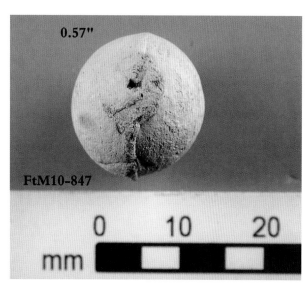

THE BASIC
MUSKET BALL

fired, the black powder ignites, causing explosive forces to expel the musket ball out of the barrel. Some sulfur melts and mixes with unburnt charcoal and saltpeter and is also pushed out of the barrel. Much of it coats the inside of the barrel and semisolidifies with a greasy texture, a process known as fouling. Every shot reduces the inside diameter of the musket, decreasing the windage between the barrel wall and the musket ball. Each musket ball becomes increasingly more difficult to ram down the barrel, requiring more time for a soldier to prepare to fire. This time can give an enemy more time to advance without being fired upon. A point is reached at which a ball will jam, or foul, partway down the barrel and the gun is no longer useful.

Numerous graduated-diameter musket ball gang molds have been found for use with civilian hunting or "fowling" muskets. Each musket ball used has a slightly smaller diameter than the previous ball. The shooter would step down the ball size after each shot. However, this is not practical for military use. The smaller the windage, the less hot gas from the powder explosion leaks past the ball and the straighter the ball will travel down the barrel, which increases accuracy. The objective of a military smoothbore musket used by a linear formation of infantry is to *get the lead out* (as the saying goes)! Accuracy is less important than the quantity of musket balls fired at the enemy.

Musket balls can be found in many different sizes. Common sizes for the American Revolutionary War period are shown in figure 2.8. Musket balls were usually categorized not by diameter but by how many musket balls were in a

2.8. Examples of musket balls with different diameters excavated at Monmouth Battlefield State Park, New Jersey.

MUSKET BALL AND SMALL SHOT IDENTIFICATION

pound. For example, a military service British Brown Bess musket had a bore of 0.75 inches, or .75 caliber, and took musket balls that were fourteen to the pound or twenty-nine per two pounds. This is equivalent to a 0.693-inch-diameter musket ball (Muller 1977: 14). A .69 caliber French military musket took a 0.63-inch-diameter ball.

By knowing a musket ball's diameter, one can estimate the bore of the gun it came from. However, what if the musket ball was fired, hit something, and is no longer round? The diameter cannot be measured directly. Another method has been devised to estimate the diameter of a nonspherical musket ball. This was demonstrated in 1780 by John Muller:

> From the specific gravity of lead, the diameter of any bullet may be found from its given weight. For since a cubic foot weighs 11325 ounces by our table, and 678 is to 355 as the cube 1728 of a foot, or 12 inches, is the content of the sphere, which therefore is 5929.7 ounces; and since spheres are as the cubes of their diameters, the weight 5929.7 is to 16 ounces, or one pound, as the cube 1728 is to the cube of the diameter of a sphere which weighs a pound; which cube therefore is 4.66263, and its root 1.6706 inches, the diameter sought. (Muller 1977: 13)

However, Muller based his calculations on the specific gravity of pure lead, which has a density of 11.338 grams per cubic centimeter, or 185.8 grams per cubic inch. He used this value to calculate back to the diameter of a lead sphere based on the weight in grams. The Sivilich Formula applies this concept to calculate the original spherical diameter of the flattened, impacted musket balls excavated at Monmouth Battlefield State Park.

Occasionally the lead was mixed with other available materials, such as tin, which may come from melting down pewter objects (80 percent tin, 20 percent lead) and mixing this with pure lead to stretch the available lead supply. Eighteenth-century lead musket balls can also contain a variety of impurities and air trapped during casting. These factors will lower the density of the overall musket ball being measured. Figure 2.9 is a musket ball that was accidentally sheared by a shovel during a metal-detecting survey at Monmouth Battlefield State Park in New Jersey. An air cavity was discovered inside the ball.

Modern competition musket balls are made by cold swaging. The Hornady Manufacturing Company is one of the top suppliers of these products, and the company's 2010 catalog states, "Cold swaging from pure lead eliminates air pockets and voids common to cast balls. And the smoother, rounder surface of a Hornady round ball assures better rotation for consistent accuracy. It's our strict production procedures which give Hornady Round Balls the unsurpassed uniformity demanded by the world's best shooters." Modern machine-made musket balls will have a density much closer than that of pure

2.9. A musket ball that was accidentally sheared by a shovel during an excavation at Monmouth Battlefield State Park. The air pocket discovered in the interior of the musket ball is visible. *(Photograph by Eric Sivilich.)*

lead, whereas cast musket balls can have a lower density because these types of imperfections.

One common mistake that is made when cleaning lead artifacts is that the patina is scrubbed off to expose the gray lead of the musket ball. This patina should not be removed. Lead bullets should be briefly rinsed with cold water to remove only surface dirt. The patina is made of lead salts, and removing the patina removes some of the original elemental lead of the projectile. These salts have lower densities than pure lead, as follows:

lead oxide:	9.38 g/cm^3
lead chloride:	6.60 g/cm^3
lead sulfate:	6.29 g/cm^3

The gram weights and corresponding diameters (in inches) of a large sampling of eighteenth-century round balls were measured from four different archaeological sites. Two sites were occupied only by Continental troops, one site was occupied only by Crown forces, and one site was a conflict area between these two groups. The musket balls used in this study were all rinsed with cold water and the patina was left intact. Musket balls that visually appeared to be made of "pewter" alloys were not included in this phase of the study. (Detailed

MUSKET BALL AND SMALL SHOT IDENTIFICATION

information on alloys is presented in chapter 8.) During the diameter-measuring phase of this project, most of the musket balls were discovered to not be precise spheres. As a result, the diameters were measured in a minimum of three different locations on each musket ball, and the average diameter was recorded. This is because of eighteenth-century molds not having true hemispherical cavities. Caution should be used when dealing with mid-nineteenth-century and later musket balls, as they were cast using iron, machine-made molds. These molds produce a much more accurate sphere. The results are shown in table 2.1.

An overall average calculated density of 10.532 grams per cubic centimeter was obtained for the four sites. It is interesting to note that the average density of the musket balls recovered at the site occupied only by British troops is slightly higher than the average densities of musket balls found at the two American-occupation sites. Apparently the musket balls recovered from the British site had little or no alloy metal mixed with the lead.

When the calculated overall average density for musket ball lead is used, along with the weight of a nonspherical musket ball in grams, the theoretical original spherical diameter can be approximated as follows:

$$\text{density} = \text{weight} / \text{volume}$$
$$\text{volume of a sphere} = \pi \times \text{diameter}^3 / 6, \text{ where } \pi = 3.14159$$
$$\text{musket ball density} = 10.5317 \text{ g} / \text{cm}^3 = 172.5831 \text{ g} / \text{in}^3$$

Therefore:

$$\text{density} = \text{weight} / (3.14159 \times \text{diameter}^3 / 6)$$
$$= \text{weight} / (0.5236 \times \text{diameter}^3)$$

$$172.5831 = \text{weight} / (0.5236 \times \text{diameter}^3)$$

$$\text{diameter}^3 = 0.01107 \times \text{weight}$$

$$\text{diameter in inches} = 0.2228 \times (\text{weight in grams})^{1/3}$$

Table 2.1. Calculated density of American Revolutionary War musket ball lead

Site	Occupying force	Density (g/cm^3)	Sample size	Standard deviation	Minimum (g/cm^3)	Maximum (g/cm^3)
Valley Forge—Washington Chapel	American	10.57923	70	0.17374	10.19764	10.93619
Monmouth advance American camp	American	10.44195	49	0.19369	9.96528	10.83713
Battle of Monmouth	British and American	10.51951	391	0.18488	9.97305	11.62043
Neuberger—retreat from Monmouth	British	10.64583	51	0.13000	10.24298	10.87315
Overall average		**10.53167**	**561**	**0.18581**		

This formula is based on significantly more data points than the original Sivilich Formula (Sivilich 1996):

$$\text{diameter in inches} = 0.223204 \times (\text{weight in grams})^{1/3}$$

However, the differences are negligible and will affect the results only in the third decimal place for most sizes of lead shot.

In a verification of the methodology, several 99.9 percent pure lead musket balls cast from a modern-made mold (see chapter 7) were used to calculate the lead density. An average value of 11.24 grams per cubic inch was calculated. An electronic balance with a weight of one decimal place was used. This would account for the slight difference noted with the published density of pure lead. Using the density for pure lead as John Muller (1977: 13) did is a good theoretical model but can create a small level of error when calculating the diameters of actual eighteenth-century musket balls (figure 2.10). Using the wrong value can alter data interpretation for borderline sizes such as small-caliber smoothbore muskets versus rifles. As stated earlier, some causes for the difference are the inclusions of air and impurities, but the major difference is that musket balls are very rarely spherical. Because balls are cast in an imperfect mold, sprue cuts, flashing, and other physical deformities can skew the actual diameter versus the calculated diameter.

2.10. Differences in calculated musket ball diameters comparing the use of the density value of pure lead versus the density value used in the Sivilich formula.

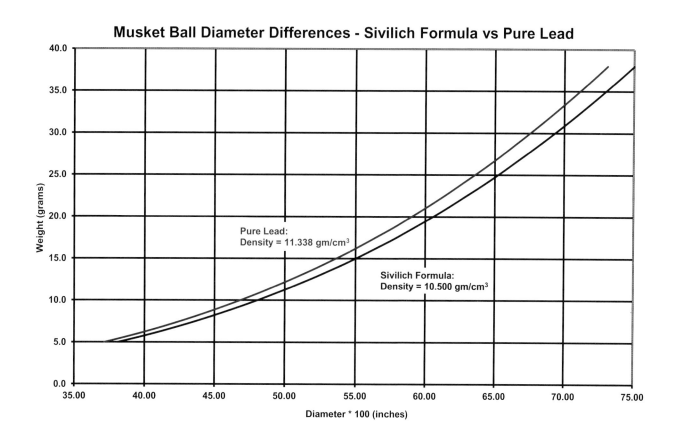

MUSKET BALL AND SMALL SHOT IDENTIFICATION

As discussed in chapter 1, research by the author and others shows that typically, but not exclusively, musket balls having diameters of less than 0.60 inches in diameter are used by rifles, musket balls with diameters between 0.60 inches and 0.66 inches are associated with a variety of smoothbore muskets such as the French-supplied "Charleville" muskets and British fusils, and lead shot with diameters greater than 0.66 inches were used in large-bore muskets such as the British Brown Bess, which was standard issue to Crown forces infantry (Goldstein and Mowbray 2010; Sivilich 2004, 2009). Small quantities of Brown Bess muskets were also used by American troops. Some were captured weapons from royal arsenals that stored older muskets used during the French and Indian War. Others were copies of Brown Bess muskets contracted by committees of safety weapons to the colonies (Neumann 1967: 22).

The general type of weapon used can be verified by analyzing musket balls from areas occupied by only one of the armies. The Continental army's advance force arrived in Englishtown, New Jersey, on June 27, 1778. The main body of the army arrived during the morning of June 28, dropped their packs, and hurried on to the battle. Most of the army returned to the camp on June 29 before marching out July 1. One of the camp sites was discovered by several local metal detectorists, who allowed me to study and photograph the artifacts they had found. Figure 2.11 is a distribution analysis of the diameters of the musket balls

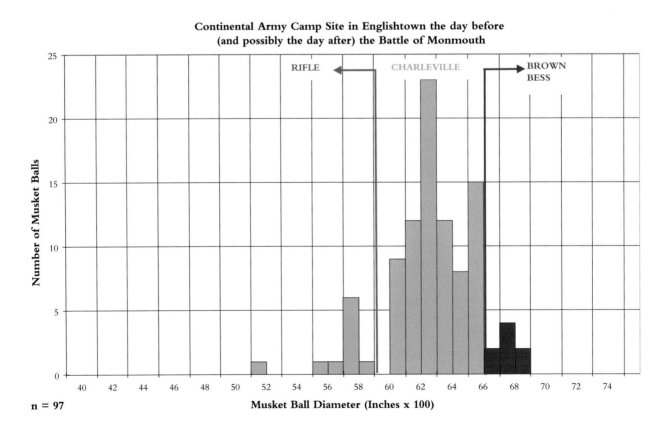

2.11. Analysis of ninety-seven musket balls recovered from an American camp site in Englishtown, New Jersey, occupied the day before (and possibly the day after) the Battle of Monmouth.

found. Nearly all of these musket balls had been dropped or discarded. The chart clearly shows that the majority of the musket balls found were 0.60–0.66 inches in diameter, with the largest concentration being at 0.63 inches. This is consistent with the size of shot used by the French-supplied Charleville muskets. The French were supplying large quantities of arms, which arrived at Valley Forge in 1777. This gave the Continental army standardized muskets and, more importantly, bayonets. At most of the battles before Valley Forge, the British would fire a few volleys and charge with fixed bayonets, scattering the American troops. An excellent example of this occurred at the Battle of Princeton, on January 3, 1777. The British Seventeenth Foot, the Fifty-Fifth Foot, and a troop of the Sixteenth Light Dragoons, all commanded by Lieutenant Colonel Charles Mawhood, pushed the Americans back with several bayonet charges. The American lines broke because they had very few bayonets (Mackenzie 2013).

Baron von Steuben trained the troops at Valley Forge how to properly use the new French muskets and bayonets in a formation that would stand up to a British charge (von Steuben 1985: 23–25, 55–56). This was seen at the Battle of Monmouth on June 28, 1778—the first battle after the Valley Forge encampment and training. Our archaeological data indicate that the Continentals held their ground and fought hand to hand against the Crown forces.

Figure 2.11 shows that there is a distinct break in musket ball diameters at less than 0.60 inches. Typically, early rifles were small-bore, and thus the musket balls of less than 0.60 inches are most likely for rifled muskets or small-bore pistols.

Another excavated site was only occupied by British troops the day after the Battle of Monmouth along the Crown forces' retreat route in Middletown, New Jersey (figure 2.12). The 0.58-inch musket ball was impacted and most likely was not associated with the encampment; it may simply have been from a farmer firing at game or a groundhog. These data clearly show a very tight cluster of musket balls between 0.68 inches and 0.71 inches in diameter. The previous graph (figure 2.11) shows a sharp break after 0.66 inches in diameter. Therefore, musket balls greater than 0.66 inches in diameter are most likely associated with larger-bore guns such as the British Brown Bess service musket.

Table 2.2 has been compiled showing the typical bore sizes for period weapons (with the ball size being 0.05 inches to 0.10 inches smaller in diameter for windage). Because of the small number of pistols that might have been used in battle, recovered musket balls are usually associated with muskets and rifles. Pistol types and bore sizes varied extensively, since many were made as personal weapons by specific gunsmiths for officers or were issued to mounted dragoons. However, the table shows a sampling of pistols that were used by the military during the American Revolution.

This data analysis is useful in associating specific ordnance sizes with a given eighteenth-century army. The size distribution of musket ball diameters,

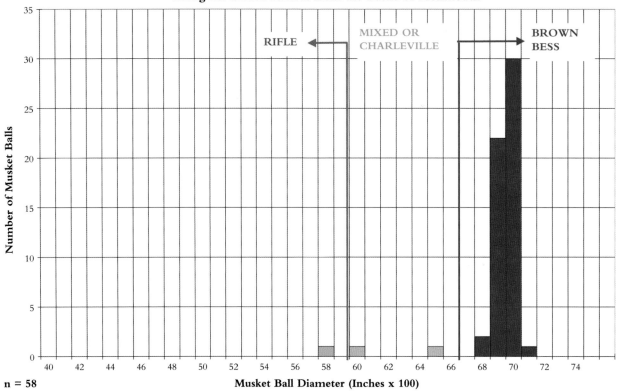

2.12. Analysis of fifty-eight musket balls recovered from the Neuberger farm, a British-occupied site in Middletown, New Jersey.

both measured for round musket balls and calculated for nonspherical musket balls, excavated to date at Monmouth Battlefield State Park is shown in figure 2.13. There are two distinct peaks, at 0.63 inches and 0.69 inches, which are typically associated with French Charleville muskets and British Brown Bess muskets, respectively.

Buck and Ball

Large-diameter musket balls could be lethal, but only if they hit their target. Because of the large amount of windage and the fact that musket balls are not perfect spheres, smoothbore muskets were low in accuracy. To compensate for this and "get the lead out," several buckshot were added to a cartridge to increase the chance of slowing down the enemy. The smoothbore musket became a "shot" gun—buckshot, that is. There is substantial evidence that shows the Americans used buck and ball during the Revolutionary War. One recruit in 1777 mentioned having "sixty-four rounds of cartridges with three buck shot in

Table 2.2. Typical eighteenth-century firearm bore sizes (or caliber)

Firearm	Bore diameter (inches)	Reference[a]
British		
Queen Anne muskets	0.75	1
Colonel's pre–Brown Bess muskets	0.80	1
Brown Bess long land muskets	0.75	1
Brown Bess short land muskets	0.75	1
Officer's fusils	0.67[b]	1
Marine and militia muskets	0.75	1
Naval muskets	0.75	1
Nock volley guns	0.46	1
Light and heavy dragoon carbines, sergeant's, artillery, Elliot	0.65	1
Wall guns	0.90–0.98	1
Ferguson rifle	0.65	1
Pattern 1776 rifle	0.615	1
Screw-barrel pistol	0.44, 0.58, 0.60	3 (pp. 152–66)
Heavy dragoon pistol	0.56	1
Light dragoon pistol	0.65	1, 3 (pp. 162–74)
Dragoon pistol	0.56, 0.60, 0.66, 0.68, 0.72, 0.75	3 (pp. 156–62)
Holster pistol	0.61, 0.62, 0.63, 0.73	3 (pp. 166–72)
Scottish regimental pistol	0.55, 0.57	3 (pp. 174–76)
Naval pistol	0.58, 0.60, 0.61	3 (pp. 208–10)
French		
Pattern 1717	0.75	1
Pattern 1728, 1746, 1754, 1766, 1768, 1772, 1774, and 1777	0.69	1
Fusils, carbines, and dragoon pistols	0.69	1
Cavalry pistol	0.60, 0.65, 0.68, 0.69	3 (pp. 180–84)
Holster pistol	0.51, 0.57, 0.58, 0.60, 0.64, 0.69	3 (pp. 180–84)
Naval pistol	0.56, 0.60, 0.68	3 (pp. 212–34)
Spanish		
Pattern 1757 infantry muskets	0.71	1
Dutch		
Generalitett	0.75	1
Pistol	0.66	3 (p. 178)
Dragoon pistol	0.69	3 (p. 178)
German		
Prussian muskets	0.75	1
Hertzberg muskets	0.77	1
Hessian Jaeger rifle	0.51, 0.60	3 (p. 136)
Hessian field Jaeger Corps rifle	0.65	4
German pistol	0.59	3 (p. 192)
Holster pistol	0.63	3 (p. 192)
Cavalry pistol	0.68	3 (p. 194)
American		
American rifle	0.50, 0.51, 0.52, 0.55, 0.57, 0.59, 0.60, 0.64, 0.65	3 (pp. 138–46)
American "Mountain" rifle	0.66[c]	3 (pp. 138–46)
Contract Committee of Safety muskets	0.70, 0.75, 0.79, 0.80, 0.90	2
Cavalry pistol	0.63, 0.65, 0.68	3 (pp. 196–98)
Holster pistol	0.56, 0.58, 0.69, 0.70	3 (pp. 200–204)
Naval pistol	0.67	3 (pp. 212–34)

[a]Reference numbers: 1, Bill Ahearn, personal communication, August 7, 2012; 2, Ahearn 2005: 148–61; 3, Neumann 1967 (page ranges shown); 4, Bailey 2002: 67–68.
[b]Diameters vary, but most are 0.67 inches.
[c]Eight grooves.

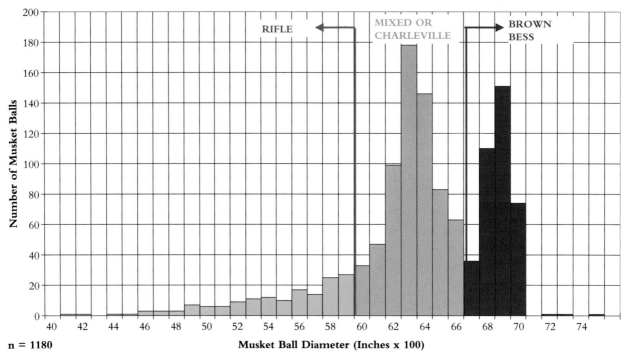

2.13. Analysis of 1,180 musket balls excavated at Monmouth Battlefield State Park.

each" (Avery 1918: 74). In 1777 Washington recommended that the men load their muskets with one musket ball and four to eight buckshot for their first volley (Peterson 1968: 60–61). In October 1777 orders, he made the use of buckshot standard practice: "Buck shot are to be put into all cartridges which shall hereafter be made" (Washington 1777–78). Solomon Parsons of the Fifteenth Massachusetts Regiment fought at the Battle of Monmouth and stated, "I beheld the red-coats within eight rods. I was loaded with a ball and six buck-shot" (quoted in Washburn 1860).

The impressions found on musket balls (figure 2.14 on page 33) indicate that the minimum load was three buckshot and one ball. The far left musket ball (artifact 2-746) was found behind the American lines where the troops rested during the largest land artillery exchange of the Revolutionary War. The musket ball in the photograph to the right (VF8-1006) was found at the Washington Memorial Chapel site at Valley Forge, Pennsylvania. This was occupied only by American troops during the winter of 1777–78.

Figure 2.15 on page 33 shows what a typical minimum load would look like and the relative size differences between buckshot and ball (buckshot is discussed in more detail in chapter 9). These artifacts were not found as a grouping but are shown simply to illustrate the sizes of typical ball and buckshot.

The theoretical packing arrangement for a 0.64-inch-diameter musket ball and three 0.30-inch-diameter buckshot in a 0.69-inch barrel (Charleville musket) is shown in figure 2.16. This is consistent with various musket balls with buckshot impressions that I have observed and measured.

THE BASIC MUSKET BALL

2.14. Three musket balls with buckshot impressions. *Left:* Two musket balls excavated at Monmouth Battlefield State Park, New Jersey. *Right:* A musket ball (0.64" in diameter) excavated at Valley Forge.

It is rare to excavate ball and buckshot together, because the paper cartridge deteriorates rapidly. Many sites have been disturbed by plowing, animal and insect burrows or bioturbation, freeze/thaw ground upheavals, and so forth. The ball and buckshot shown in figure 2.15 were found at a Revolutionary War site in a wooded area in New Jersey. The exact site location has been properly documented. However, they were not found as shown but rather were in a cluster of ball and buckshot (shown in figure 2.17 on page 34) found in one cache. In total, there are fifty musket balls and ninety-four buckshot.

The distribution of the diameters is shown in figure 2.18 on page 35. The range of musket ball sizes suggests that these artifacts were probably not in cartridge form but most likely loose in a pouch. The ratio of buck to ball is roughly 2:1, lower than expected for Continental army cartridges, which should be a minimum of 3:1.

What caused the indentations? Obviously the buckshot was pressed into the musket ball. However, the specimens shown in figure 2.14 are round/dropped musket balls. Figure 2.19 on page 35 shows pre–Civil War and Civil War–era paper cartridges (Thomas 1997: 107, 113) for round musket balls. These were used in smoothbore muskets similar to those used during the Revolutionary War. The buckshot was at the base of the cartridge, so that the

2.15. Typical example of buck and ball. These artifacts (also shown in figure 2.17) were excavated by Bob Hall at Raritan Landing, New Jersey.

MUSKET BALL
AND SMALL SHOT
IDENTIFICATION

2.16. Musket ball and three-buckshot packing diagram for an American cartridge for a Charleville musket. All dimensions are in inches. *(Drawings by Raya Lim.)*

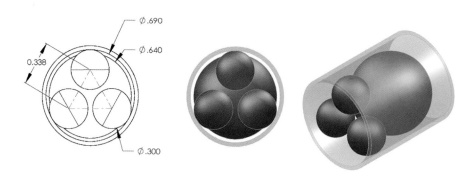

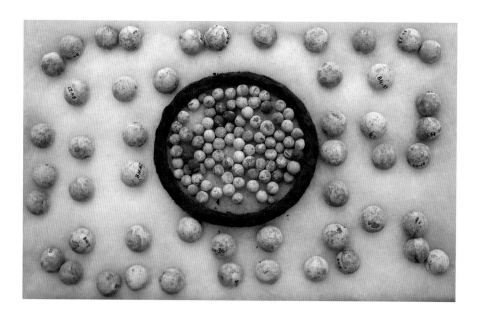

2.17. Buck and ball excavated as a cluster by Bob Hall at Raritan Landing.

loading order would have been powder, ball, and then the buckshot. The cartridges were carried in a cartridge box with a rigid cartridge holder. The cartridges would have been sitting with the buckshot against the wooden base, with the musket ball on top of the buckshot. The vibrations from walking or the pounding from running could have caused sufficient energy for the buckshot to leave impressions in the musket ball. Only three buckshot are represented in the drawings and photographs; however, this is the minimum that would have been in contact with the ball. In reality, several layers with three shot per layer could have been stacked into a cartridge.

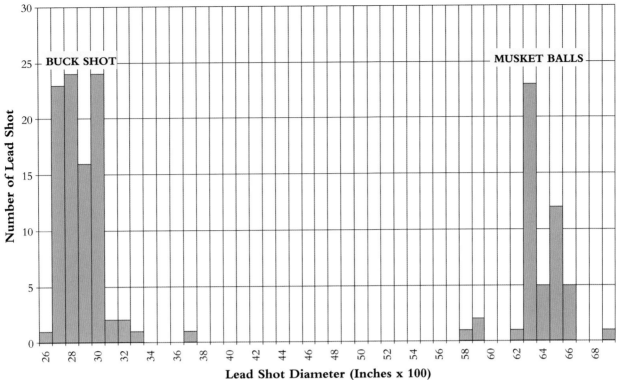

2.18. Size distribution of the buck and ball cluster excavated by Bob Hall at Raritan Landing.

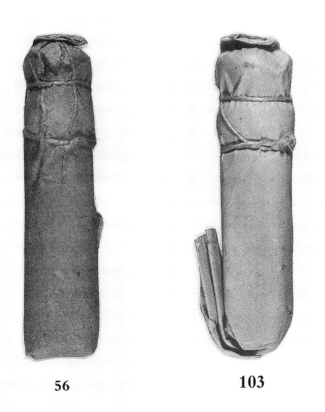

2.19. Pre–Civil War buck and ball (0.70" diameter) on the left (Thomas 1997: 107) and Civil War–era buck and ball (0.65" diameter) on the right (Thomas 1997: 113), both with the buckshot at the bottom of the cartridge. Reproduced with permission from Dean S. Thomas.

2.20. Musket balls with ramrod marks excavated at Monmouth Battlefield State Park (*left*, artifact 242-13-007) and a skirmish site from the Battle of Monmouth on Craig Road, Manalapan, New Jersey (*right*, artifact 9M9-1), which was ironically lost because of urban expansion.

Ramrod Marks

Musket balls have been excavated at Monmouth with shallow circular depressions (figure 2.20). My personal experience firing black powder flintlock muskets leads me to believe that these are ramrod marks. Even though a ball may be sitting loosely in the breech of a musket, it is still rammed down to compact the powder. If the barrel is fouled or a ball is oversized for the bore of a musket, the ball has to be rammed multiple times to seat it into the powder. If it does not seat, the air space between the powder charge and the jammed ball can compress enough upon firing to rupture the barrel, causing injury or death to the owner.

Typical Brown Bess and Charleville ramrods were iron and had slightly convex tips (as shown by the reproductions in figure 2.21). These would have made a concave circular depression in the musket ball.

Pulled Musket Balls

As described in the previous section, musket balls can jam and have to be extracted from the musket barrel. The back end of many ramrods is threaded so a steel screw designed for the purpose can be twisted into a soft lead ball to pull or extract the ball from the muzzle. Musket balls were pulled not only when they jammed but also when necessary so as to not discharge the weapon, for either noise or safety. It was not a common practice to have a loaded gun in camp.

THE BASIC
MUSKET BALL

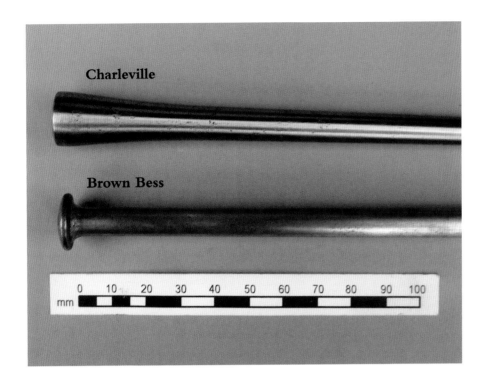

2.21. Reproduction ramrods supplied by Jim Stinson. *(Photograph by Eric Sivilich.)*

Soldiers coming off picket duty would have to clear their firelock. It was quieter and safer to pull a ball out of a barrel rather than firing it. Examples of these pulled balls have been found in numerous camp locations around the world, including Pułtusk, Poland (figure 2.22) and the United States (figure 2.23). The distinct thread of the extracting screw can be clearly seen. The lower right musket ball in figure 2.22 is a clear example of a musket ball that took more than one attempt for the screw to get a sufficient grip.

2.22. Musket balls extracted with a screw that were excavated at the site of the Battle of Pułtusk, Poland, December 26, 1806. *(Photograph by Paweł Kobek and provided by Jakub Wrzosek of the National Heritage Board of Poland.)*

MUSKET BALL
AND SMALL SHOT
IDENTIFICATION

2.23. Musket balls extracted with a screw that were excavated at American camp sites: *top left*, at Fort Montgomery State Park, New York *(photograph by Dana Linck)*; *top right*, at the U.S. Military Academy, West Point, New York; *bottom left* and *bottom right*, by Russ Balliet at Wayne's camp in Englishtown, New Jersey.

Muskets also had an attachment called a worm. This is a double helix of twisted iron terminating in two sharp points. The other end had a female thread that could be screwed onto the male-threaded end of a ramrod. Typically, worms were wrapped with a linen patch and run down a barrel to clean out the fouling. An excellent example of an actual eighteenth-century worm is shown in figure 2.24. It was excavated at the Breymann Redoubt at the 1777 Battle of Saratoga in New York.

Jim Stinson (personal communication, December 31, 2012) explained that in noncombat situations, such as picket duty, a loaded musket was usually wadded with the paper from the cartridge or a cloth patch to keep the ball and powder from falling out if the musket was inverted. To empty the musket without discharging it, the soldier would first have to remove the wadding. This could be done with a worm. While the worm was being twisted into the wadding, the musket ball could also be lightly marked. If the musket ball was loose in the breech, the musket could then be inverted to allow the ball, buckshot (if any was in the load), and powder to fall out.

Figure 2.25 shows three musket balls that were clearly marked by worms with a center screw. This style of worm appears to have been used for cleaning a barrel and extracting a musket ball at the same time, clearly an early multi-tool.

One such worm was found at the U.S. Military Academy at West Point, New York, in the early twentieth century by William Calver and Reginald Bolton (figure 2.26). It has a coarse thread screw in the center for pulling a musket ball and a double helix for grabbing the wadding. The other end was threaded to attach to the end of a ramrod. West Point was the site of Fort Putnam, from which the American forces controlled a northern section of the Hudson River during the Revolutionary War.

Several musket balls have been discovered with worm marks opposite buckshot impressions (figure 2.27). The worm marks are not deep enough to suggest that they gripped enough lead to extract the balls, which may indicate that the balls were wadded. However, the worm-marked area appears to be flattened. The buckshot impressions on the opposite side are contrary to a standard load, in which the buckshot would be in front of the musket ball and would be hit by the worm. One theory is that these musket balls were taken from ball and buck cartridges and intentionally flattened and marked with a worm as a gaming piece.

THE BASIC MUSKET BALL

2.24. British musket worm excavated at the 1777 Saratoga Battlefield. *(Collection of the Saratoga National Historical Park, SARA-1701, New York.)*

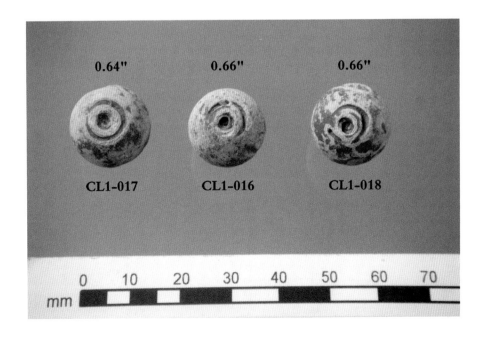

2.25. Musket balls extracted with worms that were recovered by Glen Gunther at an American camp site in New Jersey.

2.26. Musket worm excavated by William Calver and Reginald Bolton of the New York Historical Society Field Exploration Committee at Fort Number 4 at West Point. *(Collection of the New-York Historical Society, INV.5925.171.)*

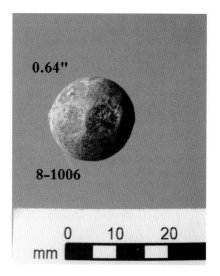

2.27. Musket balls with worm marks and buckshot impressions. The top musket ball (artifact 8-1006; three views) was excavated at the Washington Memorial Chapel site, Valley Forge. The bottom musket ball (artifact CL1-001; two views) was recovered by Glen Gunther at an American camp site in New Jersey.

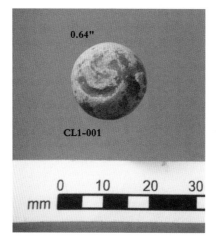
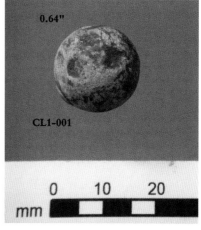

Rejected Musket Balls

Often musket balls were cast and did not mold acceptably. If a mold was cold, the lead might become stratified, making for an irregular surface. If insufficient molten lead was poured into the mold, then only a partial musket ball would be produced. The left and center photographs of figure 2.28 show a musket ball excavated at Fort Montgomery, New York, that appears to exhibit both characteristics. If the two halves of a mold did not seat properly or if the mold was not perfectly spherical, one could have a musket ball that was oval and irregular in cross section (as seen in the right photograph of figure 2.28, excavated in Valley Forge, Pennsylvania).

2.28. Rejected musket balls: *left* and *center*, two views of the same musket ball (artifact 10-939H) excavated at Fort Montgomery; *right*, a musket ball (artifact 10-255) excavated at Valley Forge.

European Musket Balls

The examples shown in the earlier section have been primarily from American Revolutionary War sites. European musket balls will now be lightly explored, but this is not to be construed as a definitive analysis. It is presented simply to show some similarities and some differences.

Numerous spherical musket balls have been found at many European conflict sites. This seems to be especially true for the British Isles. All the musket balls shown earlier attributed to British use were originally spherical. The left photograph in figure 2.29 shows musket balls from the 1690 Battle of Boyne in Ireland on display in the site museum. Shot from both Irish forces and English forces are shown. A hoard of 2,701 musket balls was uncovered in one location in Ballymore, Ireland, by a metal detectorist operating illegally. This was at the site of a Jacobite fort besieged by Williamite forces in 1691 (Shiels, personal communication, February 10, 2005).

MUSKET BALL AND SMALL SHOT IDENTIFICATION

2.29. Left: Musket balls on display at the site of the 1690 Battle of Boyne in Ireland, with those fired by Irish forces to the left and those fired by English forces to the right *(photograph by Adrian Devine). Right:* A hoard of 2,701 musket balls dug illegally at the site of the 1691 battle near Ballymore, Ireland, now in the National Museum of Ireland *(photograph by Damian Shiels).*

EXTENDED SPRUE MUSKET BALLS

Not all musket balls are spherical. Amphora-shaped lead musket shot and round musket balls with the casting sprues still attached appear to have been used in continental Europe, especially in the Baltic regions, and were also brought to America at least as early as the seventeenth century. These are known as extended sprue musket balls. The diagram in figure 2.30 shows how a musket paper

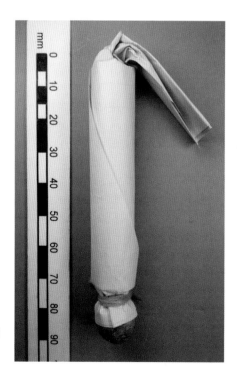

2.30. Left: A drawing of an Austrian/Hungarian military "Muskettenpatrone," or musket cartridge, showing how the extended sprue is used to secure the ball to the paper cartridge. *Right:* A reproduction paper cartridge made by the author using an authentic extended sprue musket ball from the 1655–75 Dann site, a Seneca Iroquois village.

cartridge was to be made in 1700 for the Austrian/Hungarian military (Dolleczek 1896). It shows that the end of the paper cartridge was gathered around the uncut casting sprue and string was tied around the paper and sprue to keep the musket ball secure. The photograph in figure 2.30 is a reproduction cartridge using an original seventeenth-century extended sprue musket ball (ball 12 in figure 2.37). Other countries may have had different styles of cartridges, but that is beyond the intent and scope of this book.

These types of shot could be more damaging to tissue than a round ball. If the shot hit broadside rather than the rounded tip first, it would tear the flesh, making an irregular hole. A round musket ball hole was much easier to deal with medically.

Adrian Mandzy of Morehead State University in Kentucky has been conducting archaeological surveys at battle sites from a range of time periods in the Ukraine and Poland. Figure 2.31 is a group of musket balls that his team of graduate students excavated at the site of the 1649 Battle of Zboriv, Ukraine. A mix of types and sizes were found, including several extended sprue examples.

Figure 2.32 shows musket balls excavated by Bo Knarrström from the 1677 Battle of Landskrona in Sweden. Again there is a mix of extended sprue ordinance with spherical musket balls (Knarrström 2006: 63, 67).

2.31. Lead musket ordnance excavated at the site of the 1649 Battle of Zboriv, Ukraine. *(Photograph by Adrian Mandzy, PhD.)*

 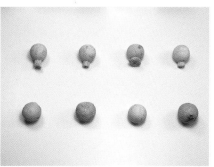

2.32. Lead musket ordnance excavated at the site of the 1677 Battle of Landskrona, Sweden. *(Photographs by Bo Knarrström, PhD.)*

MUSKET BALL
AND SMALL SHOT
IDENTIFICATION

2.33. Left: Musket ball with an extended sprue excavated at the site of the Battle of Pułtusk, Poland, December 26, 1806 *(photograph by Paweł Kobek and provided by Jakub Wrzosek of the National Heritage Board of Poland). Right:* Reproduction lead musket ball with the casting sprue intact produced by Andy Drysdale.

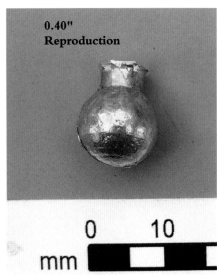

2.34. The caption at the antiquities website where this image originally appeared read "Circa 1640's. Original lead shot or 'musket balls' used by matchlock type muskets and pistols. Reference: Bailey, G. Finds Identified, p. 69–72. Very fine condition. Provenance: found on a Charles I Civil War battle site near Harston, Cambridgeshire." *(Image courtesy of Time-line Originals—http://www.time-lines.co.uk/.)*

Extended sprue ordnance was not used just in the seventeenth century. The left photograph in figure 2.33 is a musket ball excavated at the 1806 Battle of Pułtusk, Poland. This is compared to the photograph on the right of a 0.40-inch musket ball cast by Andy Drysdale, a friend of mine and fellow reenactor, on which the casting sprue was not cut off. This clearly demonstrates how these types of shot were produced. The lead was cast in a mold and the sprues were simply left on. Some sprues appear to have a small cylindrical hole in the center (as shown in later examples). Whether these were formed during the casting process or mechanically made when the lead was still molten is unclear.

Musket balls with casting sprues have also been reported to be found in the United Kingdom (fig. 2.34), according to Glenn Foard (personal communication, 2014) of Huddersfield University, who has been excavating the 1640s English Civil War.

A possible variation of this type of "musket ball" is shown in figure 2.35. Round balls with extended sprues appear to have been cold hammered into jug-shaped shot. The left photograph is of an artifact excavated at the site of the 1709 battle in Poltava, Poland. The right artifact, found in Cheshire, England, in 2008, has a weight of 22.3 grams and a diameter of 14.71 millimeters (0.579 inches). The precise context of this artifact was not given. The purpose for this alteration is not known but is presumed to have been to fit the barrel of a musket with a smaller musket ball than what was immediately available (Mandzy 2012: 74).

Extended sprue musket balls were transported from Europe to the colonies. They were traded to local Native Americans by the British, French, and

THE BASIC
MUSKET BALL

Dutch. Figure 2.36 shows photographs of European-style extended sprue musket balls found at seventeenth-century Seneca village sites in upstate New York, now displayed in the 1655–75 Dann, Marsh, and Wheeler Village Sites exhibit of the Rochester Museum and Science Center in Rochester, New York. The center photograph is a close-up of the right musket ball shown in the left photograph, showing the distinctive lead cap at the end of the casting sprue, which could be used as a tie point for the paper cartridge. The musket balls shown in the right photograph of figure 2.36 are displayed in the Rochester Junction and Boughton Hill Village Sites exhibit at the Rochester Museum. These artifacts were excavated by members of the New York State Archaeological Association in the mid-twentieth century.

Adrian Mandzy (personal communication, February 10, 2013) indicated that in the early 1980s he interned at the Rochester Museum and was familiar with the collections. He recalled that an analysis of musket locks found at these sites (Puype 1985) indicated that they were mostly Dutch.

At the time of this writing, a group of extended sprue musket balls from the Dann site came up for sale on eBay by Michael Albanese, who had acquired them at an estate sale for the late William Carter. They were listed as being

2.35. Left: Jug-shaped musket shot from the Battle of Poltava, Poland *(photograph by Adrian Mandzy, PhD)*. *Right:* Postmedieval jug-shaped musket ball found in Cheshire, England *(photograph courtesy of the Portable Antiquities Scheme website, www.finds.org.uk, object no. LVPL-8F29F8).*

from the personal collection of Mr. William Carter with his site identification tag still attached to the ring. The Dann Seneca Iroquois village site was located in Honeoye Falls, New York and dated from 1655–1675 and covered 12.5 acres with an estimated population of 1,700–2,800 Natives.

. . . part of an estate sale from the estate of Mr. William Carter a very early Western New York Historian and collector. Mr. Carter started collecting well before the 1950's. . . . Mr. Carter was an early member of the respected Morgan Chapter of New York artifacts [*sic*] historians and collectors.

2.36. Artifacts excavated at several 1655–75 Seneca village sites in upstate New York. All items are in the collections of the Rochester Museum and Science Center, Rochester, New York.

MUSKET BALL
AND SMALL SHOT
IDENTIFICATION

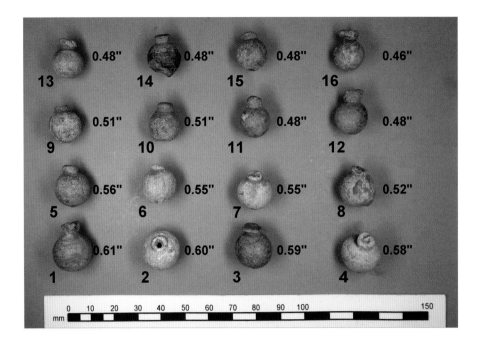

2.37. Extended sprue musket balls excavated at the Dann site, a Seneca Iroquois village in Honeoye Falls, New York, and purchased by the author.

I purchased the collection of sixteen extended sprue musket balls to study early bullets in America (figure 2.37).

Other seventeenth-century sites in this country have not produced evidence of extended sprue musket balls being used. Saint Mary's City, Maryland, was a British colony established in 1634, and it was active only until 1695 (Historic St. Mary's City 2013). It is a seventeenth-century time capsule. Numerous standard cut-sprue musket balls and small shot have been excavated at the St. John's, Town Center, and Pope's Fort sites, but none of these had extended sprues (Henry M. Miller, personal communication, May 7, 2013).

A 0.44-inch-diameter extended sprue musket ball was excavated at Monticello (Thomas Jefferson's home) in Charlottesville, Virginia, in the same context as a Jeffersonian doorknob (Michelle Sivilich, personal communication, February 14 and 20, 2013). On the Monticello Internet website (www.monticello.org/site/research-and-collections/firearms), Jefferson is described as an avid gun collector and a fair marksman. He brought back guns from France and had a pair of Turkish pistols that were given to him from the estate of General Isaac Zane in lieu of money.

CHAPTER 3

WHAT DID IT HIT?

MUSKET BALLS WERE DESIGNED FOR A SPECIFIC PURPOSE: to be fired from a weapon at a specific target with deadly intent.

Because of the windage between the ball and the musket barrel wall, the accuracy of a smoothbore gun was low (as compared to rifles) from 100 yards and beyond, and the musket ball did not always hit the intended target. British Colonel George Hanger described this: "A soldier's musket, if not exceedingly ill bored and very crooked, as any are, will strike the figure of a man at 80 yards; it may even at a hundred; but a soldier must be very unfortunate indeed who shall be wounded by a common musket at 150 yards, provided his antagonist aims at him, and, as to firing at a man at 200 yards with a common musket, you may just as well fire at the moon and have the same hopes of hitting your object" (Hanger 1814: 205). Test firings to determine the accuracy of smoothbore muskets were conducted by Joseph Bilby as research for *Monmouth Court House: The Battle That Made the American Army* (Bilby and Jenkins 2010). He concluded that muskets fired with just a musket ball were very accurate at 50 yards, but when buckshot was added, the accuracy decreased (Bilby, personal communication 2013; Bilby and Jenkins 2010: 177–79).

MUSKET BALL
AND SMALL SHOT
IDENTIFICATION

So what did a musket ball hit? Usually a soft lead musket ball will exhibit various forms of deformation depending on what it hit. My aim with this chapter is to help the reader identify what a projectile may have hit, either intentionally or accidentally, by examining a number of different characteristics of impact deformation.

Soft Target Hit

When a musket ball misses the target and hits a soft or compressible material such as clean dirt, sand, water, and so forth, then it will exhibit only minor deformation (such as those deformations shown in figure 3.1). This is especially true if the firelock is aimed high and the ball does not hit an enemy soldier but instead continues until it loses velocity and hits the ground at a relatively slow speed. The force of impact is very low (force equals mass multiplied by acceleration), and the musket ball tends to skip a few times and roll to a stop with minimal deformation. These observations are based on experimental firings at Monmouth Battlefield State Park in April 1992.

"Barrel Band"

Some musket balls that have hit soft ground (such as ball 90D10-1 in figure 3.1) may still appear to be round, which complicates the determination of whether it had been fired. Sometimes there is subtle evidence on the ball indicating that it was fired. If a musket ball fits tightly in a barrel and was rammed down the barrel without rolling, the musket ball can have a flat band around all or part of its circumference from the bore of the gun; this is known as a "barrel band" (not to be confused with the metal band on the musket that holds the barrel in place).

3.1. Three musket balls that struck soft materials. These were excavated in a sandy soil area of Monmouth Battlefield State Park, New Jersey.

  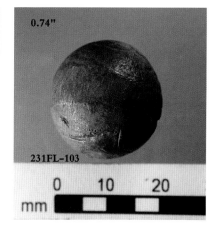

Artifact 224-12-026 (figure 3.2, *left* and *center*) is an extreme case of barrel banding, which is useful to show the detail in a photograph. This ball shows no evidence of rifling or of being patched. Therefore, it was probably fired by an American soldier with a musket having a bore size of 0.58 inches, since that is the measured diameter of the musket ball at the band. Either the ball was sized too close to the bore size or the musket was fouled, because the ball had to be rammed with enough force to create a deep ramrod impression (figure 3.2, *center*). The ball was most likely a bit large, since fouling is a soft, greasy substance that would not leave a physical impression on the ball. The reproduction ball in the right photograph demonstrates this.

A firelock experiment was held at Monmouth Battlefield State Park to determine accuracy at a specific distance and elevation. Reproduction Brown Bess muskets with 0.75-inch bores were used for the test. After the test firings, members of BRAVO used metal detectors to recover the musket balls. Typically, reenactors will use very tight-fitting musket balls for competition matches to reduce windage and improve accuracy. During this experiment, ball 231FL-103 was originally 0.74 inches in diameter. The tight fit caused a "barrel band" impression on the musket ball (figure 3.2, *right*).

3.2. Musket balls with "barrel bands." Artifact 224-12-026 (*left* and *center*) was excavated at Monmouth Battlefield State Park. Ball 231FL-103 (*right*) is a reproduction fired from a 0.75" bore Brown Bess musket and recovered as part of an experiment at Monmouth Battlefield State Park.

Tree Impacts

Musket balls often hit trees. If a soldier was close enough, the musket ball would embed in the tree. John Mills of Princeton Battlefield State Park conducted an experiment to show the force of impact in a tree for the park museum (figure 3.3). Mills used a reproduction Brown Bess musket for this test and fired it at close range at the tree. There is little deformation to the musket ball, because it was made from scrap lead and is most likely not pure. Modern lead typically contains antimony to make it harder.

MUSKET BALL AND SMALL SHOT IDENTIFICATION

How do these types of musket balls end up in the ground at a site? Some trees may be attacked by insects, such as carpenter ants, which can hollow out the interior of the tree. In other cases the tree may have fallen and rotted, after which the impacted musket ball would have simply fallen out.

Still other trees with musket balls in them may have been cut down and removed from the site. A modern example of this occurred at Monmouth Battlefield State Park. To improve the viewing area for a new interpretative center, some trees had to be cut down. These trees had been in the impact zone for numerous reenactor live-round firelock matches at the park. The park's maintenance staff occasionally would hit musket balls with chainsaws when cutting down trees in the firelock impact area.

Since the 1980s, live-round competition firelock matches have been held at Monmouth Battlefield State Park in an area that did not have any 1778 musket activity. Numerous reproduction musket balls were fired on a downhill slope at target boards, which they easily passed through and would then embed in the dirt or trees beyond. Some musket balls literally bounced off trees, while others (such as the one in figure 3.3) would embed. Over the decades, the trees healed with the reproduction musket balls inside. Annually, members of BRAVO assist Garry Wheeler Stone, the park historian, in demonstrating electronic archaeological techniques to visiting school groups. Reproduction balls are located with metal detectors, and the students help dig up the objects and roughly map their positions. The firelock impact area is a valuable source for reproduction musket balls; these can be used for comparative analysis with actual musket balls found at conflict sites. Occasionally fragments of reproduction musket balls have been found (as shown in the left side of the left photograph in figure 3.4). The striations in the lead suggest that these were cut with a chainsaw by a maintenance staff member.

To confirm this hypothesis, I conducted an experiment with the assistance of Colin Parkman, a friend and graduate student from the University of Bradford in Yorkshire, England. A 0.69-inch-diameter reproduction musket ball

3.3. Reproduction musket ball fired into a tree by John Mills of Princeton Battlefield State Park, New Jersey, and recovered for display in the park museum.

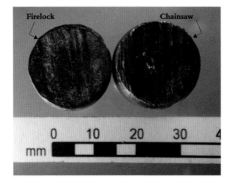

3.4. Half of a reproduction firelock musket ball recovered at Monmouth Battlefield State Park (*left bisected ball in left photograph*) and a reproduction musket ball intentionally bisected with a chainsaw (*right bisected ball in left photograph* and *right photograph*) as an experiment by the author and Colin Parkman.

was embedded between two 2×4's, and the assembly was cut with a chainsaw (figure 3.4, right photograph). The results matched, as shown in the left photograph in figure 3.4. If you have a cut musket ball with this type of pattern, the cut marks are from the twentieth century! The key word is "cut." The musket ball's calculated diameter based on the weight as compared to the measured or approximated diameter can determine whether the ball has a lead loss or was simply flattened by impact on a grainy surface (see the section "Fence Rail Impact" below).

If a musket ball was slowing down after reaching maximum velocity and hit a tree, often it would not embed in the wood but would bounce off and drop to the ground. Because of its high mass, the force would be great enough to deform the soft lead into a hemisphere, as shown with the two examples in figure 3.5. Numerous hemispherical-shaped musket balls have been excavated by BRAVO at Monmouth Battlefield State Park. Many were in areas known to have been orchards or woodlots at the time of the battle. The two examples shown were recovered in the area of the Battle at the Hedgerow, a segment of the June 28, 1778, Battle of Monmouth. During this phase, British troops pushed American forces back to the American artillery line, but intense musket and artillery fire caused heavy British casualties. The large number of impacted/hemispherical musket balls excavated in this area suggests that they may have hit some trees that were in 1778 a "hedgerow" or a green wood rail of the hedged fence. A 1778 map by Michel Capitaine du Chesnoy, cartographer to General Lafayette, does not illustrate this area as being an orchard or woodlot (Capitaine 1778); however, a few trees may have been present.

WHAT DID IT HIT?

3.5. Two musket balls that appear to have hit trees but probably did not penetrate the wood. The center and right photographs are different views of the same musket ball excavated from area known as the "Battle at the Hedgerow," part of the Battle of Monmouth.

3.6. Reproduction musket balls in various states of deformation that hit trees but did not lodge in the wood.

As an experimental confirmation, figure 3.6 shows three reproduction musket balls recovered at Monmouth Battlefield State Park in the area that had annual twentieth-century live-round competition firelock matches for many years. Fortunately this area has no battle-related archaeological context. Therefore, all of the shot recovered there was modern. The musket balls were found beyond the particleboard target frames, clustered at the base of trees approximately 125–150 yards from the shooters. Slowed down by impact with the particleboard, some of these musket balls apparently did not lodge in the trees but deformed on impact, then apparently bounced off and fell to the ground.

Many musket balls with hemispherical shapes and relatively smooth impact surfaces have been recovered in combat zones reported to be orchards or woodlots. They are useful for defining areas that once had trees but not shown in the historical records. However, this type of deformation may also be possible from hitting another kind of object, such as a wooden structure, at a low velocity.

If the velocity of a musket ball is high enough, the ball can embed in the bark and partially into the wood. Over time, the ball can fall out of the tree for many different reasons, such as those given earlier. There are distinct characteristics that can be seen in the deformed lead, as shown in figure 3.7. The part of the musket ball that contacts the tree is usually convex and has linear gouges

3.7. Musket ball (two sides shown) that hit and lodged in a tree at the Battle at the Hedgerow conflict area at Monmouth Battlefield State Park.

from the lead impacting the wood (figure 3.7, *left*). The opposite side typically flattens from the lead's compressing, and some, or all, of the edge may roll back as the flattened projectile enters the tree and the lead thins around the perimeter (figure 3.7, *right*).

Figure 3.8 is a musket ball that appears to have lodged in a tree. It was also found at the hedgerow at Monmouth Battlefield State Park. It has a calculated original diameter of 0.63 inches, which is typical for an American musket ball, probably from a Charleville musket. It has the typical characteristics of a tree hit as described earlier; however, it also has two circular indentations in the side that would have been facing out of the tree. These appear to be from its having been struck by smaller-caliber musket balls, possibly from rifles or pistols. One can only speculate that a British officer was behind a tree and American soldiers took most deliberate aim at him. Shooting at officers during the American Revolution was not considered to be acceptable by the British command. The Americans thought differently.

3.8. Musket ball that hit and lodged in a tree at the hedgerow at Monmouth Battlefield State Park and was then hit by two smaller projectiles.

Ricochet

Some musket balls simply hit the side of a tree or glance off of a rock, an action commonly referred to as a ricochet. Musket balls that ricochet have very distinct shapes, as shown in three examples in figure 3.9. They all have a flat bottom, as the lead is swept into a thinned tail as it contacts and slides over the impact surface (provided the velocity is high enough). This teardrop shape shows the direction the ball was traveling when it hit, as the tail is the trailing edge. The right ball in figure 3.9 is the flat side of the ricochet, and the scrapes made by the impact surface are easily identifiable, suggesting that the ball glanced off of a rough surface such as a tree. All three artifacts in figure 3.9 were found in areas known to have trees but very few large rocks at the time of the battle. Calculating the diameter based on the artifact's weight using the Sivilich

3.9. Three musket balls excavated at Monmouth Battlefield State Park that ricocheted off solid objects—probably trees.

MUSKET BALL AND SMALL SHOT IDENTIFICATION

Formula may not give very accurate results for ricocheted musket balls, since there is an opportunity for lead loss as the ball swipes across the target surface.

Lead loss is most prevalent when a musket ball hits an object that is harder than lead, such as a rock. Many examples have been excavated at Fort Montgomery in New York. This fort was built by the American army on a rock outcropping in the Hudson Highlands overlooking the confluence of Popolopen Creek with the Hudson River. It was an excellent vantage point where the Hudson narrows and curves. It had several cannon, including six 32-pounders that could check British ship movements. It was undermanned and not been completed when the British unexpectedly attacked from the rear, or north end, of the fort on October 6, 1777. After much fighting, the British overran the fort (Fisher et al. 2004). Archaeological excavations over the years have recovered numerous musket balls including ricochets, as shown in figure 3.10. They are severely deformed from ricocheting off rocks, and it is obvious that they have lost some mass because of damage. These musket balls have the distinct trailing-edge tail but can also be identified by the deep scrape marks made by the rocks. Many small pieces of lead excavated here are presumed to be pieces of musket balls.

Fence Rail Impact

Identifying what a musket ball hit is occasionally a matter of matching impressions in the lead surface with possible known objects that may have been the impact target. Figure 3.11 is of a musket ball that appears to have hit a fence rail at the Battle of Monmouth. This musket ball was found east of the Parsonage Farm apple orchard just over the boundary with the adjoining Rhea farm. In the last action of the battle, General Anthony Wayne and the Third Pennsylvania

3.10. Two views each of two musket balls excavated at Fort Montgomery State Park, New York, that ricocheted off rocks. *(Left two photographs by the author, right two photographs by Dana Linck.)*

3.11. Musket ball excavated at Monmouth Battlefield State Park that appears to have hit a weathered split rail fence.

Regiment were pushed back into the orchard by British grenadiers and the Thirty-Third Regiment of Foot. The location where it was found suggests that this musket ball was fired at the British by a Pennsylvania soldier who was in the orchard. The impact side of the musket ball has a well-defined pattern of ridges.

The question now becomes: What did it hit? A variety of materials, such as fabric and wood, were compared to the ridges embossed into the musket ball. Coarse, aged, weathered hardwood fits the pattern very well. This is consistent with the wood of a split rail fence. At present, a reproduction split rail fence is at the location where this musket ball was found. It identifies the historical boundary between the two eighteenth-century farms where this segment of the battle took place. A possible hedged fence is shown on the 1778 Capitaine map (Capitaine 1778). No other structures are shown on this map where the musket ball was found. A section of this weathered locust fence rail is shown in the right photograph with the musket ball. The ridges in the musket ball appear to match the grain of the wood. This artifact tends to confirm that the partially hedged fence between the Parsonage and Rhea farms actually existed and was constructed of hardwood (probably locust) split rails.

Impact with a Smooth Flat Object

Figure 3.12 shows a musket ball excavated at Monmouth Battlefield State Park at the battle at the Parsonage Farm. The musket ball diameter suggests that it was fired by an American soldier, possibly from a Charleville musket. The area where it was found had several outbuildings near the main house. The musket ball shows no evidence of lead loss but obviously hit a solid, hard, smooth or flat object. The

3.12. Musket ball excavated at Monmouth Battlefield State Park that hit a smooth, hard, flat object.

very smooth surface and lack of any curvature in the impact area of the ball suggest that it may have hit metal. This could have been any number of farm items such as a hinge, a shovel, a plowshare, and so forth. One can only speculate.

Impact with a Musket Barrel

There are many war stories about objects such as watches, coins, books, and other items that saved a soldier's life by absorbing or deflecting bullets that might have otherwise been fatal to the recipient. During a firefight, this may not be as rare an instance as one might imagine. Figure 3.13 shows photographs of three musket balls that all have very smooth, concave deformations. The extreme levels of deformations suggest that the balls hit very solid, cylindrical objects at reasonably high velocities. The radii of the depressions are consistent with the outside diameters of musket barrels. They were all recovered at Monmouth Battlefield State Park. Artifact 242-1-754 (figure 3.13, *left*) was found at the Battle at the Hedgerow. The large size of the musket ball (0.70 inches) suggests that it was from a British Brown Bess and thus may have been intended for an American soldier but hit the barrel of his gun instead. The calculated diameters of the center and right musket balls in figure 3.13 suggest that they were fired by Americans.

Artifact 234-9-859 (figure 3.13, *center*) was found in an area where the archaeological data place a British Highland regiment making a stand against seasoned Continental troops (Sivilich 2005; Sivilich and Stone 2009; Stone, Sivilich, and Lender 1996). In the early segments of the battle, Brigadier General Charles Scott led an advance battalion of handpicked marksmen toward

3.13. Three musket balls excavated at Monmouth Battlefield State Park that probably hit musket barrels.

Monmouth Courthouse (in Freehold, New Jersey). Upon their return from town and heading west, they encountered a force of British troops who were marching north. The Forty-Second Regiment of Foot pursued the Americans, and several running skirmishes took place. The Continentals took cover in an orchard on the Derrick Sutfin farm but were displaced by the Forty-Second. The Highlanders ended up trapped in this orchard by heavy fire from artillery that ripped through the British ranks with case and canister shot along with sharp fire from American snipers.

The Highlanders abandoned the orchard, moved south along a hedgerow, and crossed a fence line into an adjoining meadow. They were pursued by a platoon of handpicked Continentals. The Americans took a position at the fence and the Highlanders took a stand in the meadow. A volley or two were exchanged before the Forty-Second filed off. These details, described in the memoirs of Private Joseph Plumb Martin from Connecticut (Martin 1988: 128–30), have been confirmed by extensive archaeological work done by BRAVO. Artifact 234-9-859 (figure 3.13, *center*) is 0.61 inches in diameter and was found at the site of the British defense line in the meadow. The historical accounts and archaeological data suggest that this musket ball was fired from the fence line by one of the Americans and hit a British musket barrel, because the outside radius of a Brown Bess musket matches the curvature of the musket ball.

Artifact 227DS-9 (figure 3.13, *right*) was found at the Point of Woods, a very intense conflict area between the Continentals and British grenadiers during the Battle of Monmouth. This musket ball nearly wrapped around a smooth cylinder probably made of metal. There were no known structures at the site in 1778. The most probable object that it would have hit is a musket barrel.

Human Hit

Determining whether a musket ball found on a battlefield had hit and passed through a person is very difficult. The type of deformation could range from none, if it passed only through flesh, to various levels of deformation depending on the velocity of the ball and which bone or bones it may have hit, passed through, or ricocheted off. What might appear to be a deformation from hitting a tree could, in fact, be from hitting a human bone. Research work by others in the field of battlefield archaeology is just beginning to determine whether human blood can be accurately detected on lead bullets (Hartgen Archeological Associates 2008). To date, only a few musket balls have been found inside human remains at burial sites. However, there is a large collection of musket balls embedded in various human bones from the Battle of Waterloo, and these are on display at the Royal College of Surgeons, Surgeons Hall Museums in Edinburgh, Scotland (figure 3.14).

MUSKET BALL
AND SMALL SHOT
IDENTIFICATION

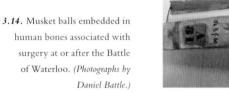
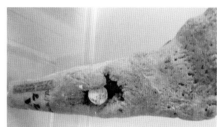

3.14. Musket balls embedded in human bones associated with surgery at or after the Battle of Waterloo. *(Photographs by Daniel Battle.)*

The top left photograph shows a musket ball lodged in a human vertebra. The top right photograph shows a musket ball in what appears to be a human tibia. The bottom left photograph shows a musket ball in a human tibia that is labeled "A31" with the description "Bones of the knee-joint, removed by amputation twenty-seven days after a gunshot injury. The ball entered the front inner aspect of the tibia exiting through the head of the fibula. The articular surfaces have been destroyed by suppurative arthritis." The bottom right photograph shows a musket ball in a human tibia that is labeled "A77" and is described as "The upper two thirds of the tibia of an artilleryman who received a musket ball into the substance of the bone without further fracture." The impact of these balls caused severe trauma, but surprisingly, there is very little distortion of the lead balls. This makes it very difficult to determine whether a musket ball excavated at a conflict site had actually hit a human target.

However, there are rare occasions when impact with a human target becomes gruesomely obvious. The musket ball shown in figure 3.15 was found at Monmouth Battlefield where heavy fighting and casualties occurred in a light woodlot. The Americans maintained a line of defense against British grenadiers to allow Washington time to set up an artillery line in the rear. Paul Kovalski, Jr., DMD, who has over twenty years of experience as a forensic dental consultant for the Monmouth County, New Jersey, Medical Examiner's Office, studied the rectangular depression in the musket ball and concluded that it appears to be that of a human front incisor. The compression of the lead indicates that the ball apparently hit a soldier in the tooth and proceeded through the cranium. The calculated original diameter of the ball was 0.63 inches, which suggests that it was

3.15. Fatal musket ball with front tooth impression excavated at Monmouth Battlefield.

fired by an American, instantly killing a British soldier. This musket ball exemplifies the realities of war.

Surgically Extracted Musket Ball

The musket ball shown in figure 3.16 (*left* and *center*) has several scrape marks that radiate in toward the center of the ball, creating a ring or small circular lip (visible in the left photograph). These markings do not appear to have been made by an animal chewing on it; an animal would not leave a ring on the ball, as the teeth marks usually continue completely to the center (see chapter 7). The markings on artifact 293-4-774 are more consistent with surgical forceps attempting to grab the ball but not getting a strong enough grip and slipping off, creating the raised ring effect. Perpendicular to this ring are several opposing deep gouges, shown in the center photograph. This would occur if the surgeon pushed the forceps deep into the wound cavity to finally grip the ball, as demonstrated by the reproduction in the right photograph. This musket ball was excavated at Old Tennent Church in Manalapan, New Jersey, the site of a Continental army field hospital: "After the action in our part of the army had ceased, I went to a well, a few rods off, to get some water. Here I found the wounded captain, mentioned before, lying on the ground and begging his sergeant, who pretended to

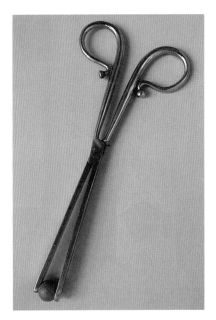

3.16. Two views of a large-diameter musket ball (artifact 293-4-774) excavated at Old Tennent Church in Manalapan, New Jersey, that appears to have been surgically extracted. The right photograph is of a reproduction ball extractor.

MUSKET BALL AND SMALL SHOT IDENTIFICATION

have care of him, to help him off the field or he should bleed to death. . . . I then offered to assist them to a meetinghouse a short distance off, where the rest of the wounded men and surgeons were. . . . I helped him to the place, and tarried a few minutes to see the wounded and two or three limbs amputated" (Martin 1988: 131).

Pulled, Reloaded, and Fired

Occasionally a musket ball type is found that is difficult to interpret. Artifact 242-9-990 (figure 3.17) is one example. It has a hole from an extractor screw. It was also fired, which is evident from the partially collapsed screw hole, the swept-back lead, and the deep concave impression. But what caused this concave impression? Based on the way the lead is swept back over the impression, it appears that something forced it either into or out of the barrel.

This is not likely to be a ramrod mark. I conducted an experiment to see whether I could make an impression as deep as that in artifact 242-9-990 with a ramrod. I manufactured a simulated ramrod tip by rounding the end of a half-inch-diameter aluminum rod. Using a three-pound sledgehammer, I struck the reproduction musket balls with the simulated ramrod, hitting each with several levels of force. (It should be noted that the reproduction balls were cast using modern lead that probably contained antimony to harden it.) The left photograph of figure 3.18 is a reproduction musket ball that was struck with an estimated force typically required to ram a tight fitting ball into a barrel. Even the heaviest blow, shown in the right photograph of figure 3.18, failed to reproduce the depth of the indentation in artifact 242-9-990. It is my opinion that this blow far exceeded the force that could be obtained by slamming a ramrod into a musket ball.

So what caused such a deep concavity in artifact 242-9-990? This ball could possibly have been "pushed" out of the barrel by another musket ball. This

3.17. Three views of a fired musket ball that also has an extractor screw hole. It might have been the leading ball of a double load. It was excavated at Monmouth Battlefield State Park in the hedgerow conflict area.

3.18. Reproduction musket balls struck with a reproduction ramrod: *left*, approximately a normal amount of force; *right*, an excessive amount of force.

scenario suggests a double load. The second ball may have had less frictional resistance in the barrel, possibly having a slightly smaller diameter. The original diameter of the leading ball was 0.69 inches, suggesting a British Brown Bess or a large-bore American gun. There is at present no way of determining whether ball 242-9-990 was extracted and reloaded or whether an unsuccessful attempt was made to extract it and it was fired out. Many other theories are possible, but a reasonable conclusion is that this musket ball was fired after being altered by an extractor.

Figure 3.19 shows a reproduction musket ball found in a firelock area at Monmouth Battlefield State Park. It clearly has an extractor screw hole (as seen in the left photograph). This suggests that it may have jammed in a barrel and had to be pulled. The corresponding ramrod indentations shown in the right photograph show that the ramrod did not make very deep impressions, certainly not to the depth seen in artifact 242-9-990.

3.19. Two views of the same fired reproduction musket ball with an extractor screw hole.

MUSKET BALL
AND SMALL SHOT
IDENTIFICATION

What Did It Hit?

The deformed shape of an impacted musket ball will vary with as the many different types of possible objects that it might strike. Making a positive determination in these instances can be difficult and can be done only by utilizing all of the available data. More often the results of the analysis will be subjective rather than conclusive. Artifact 211-7-189 (figure 3.20) is an excellent example of such an analysis. This musket ball is 0.61 inches in diameter and has an irregular grayish patina, which indicates that it has pewter mixed with the lead. This is a practice usually associated with American-made musket balls. It was excavated at Monmouth Battlefield State Park in an area identified through archaeology and historical research where a group of handpicked Continental troops were being pursued by the famous Forty-Second Regiment of Foot. The number of musket balls found and their types indicate that a firefight took place at this site. The ball size and composition suggest that this musket ball was fired by an American at the Highlanders. The musket ball has a smooth, channel-like, deep impression with a very flat bottom. It was found in a farm field that is still in agricultural use, so it could possibly have been struck by a farming implement. However, the shape, depth, and cross-sectional curvature of the impression do not appear to match the edges of typical plow blades and harrowing discs. Therefore, the possibility exists that it struck a solid object after being fired. The smoothness of the impact area suggests that it hit metal. When metal objects that may have been carried by the men in the Forty-Second are examined, one fits the pattern very well: the guard of a Scottish "basket hilt" broadsword (accurate reproduction shown in figure 3.20, right photograph). Therefore,

3.20. A musket ball with a channel-like deep impression recovered at Monmouth Battlefield State Park in an area where Scottish Highlanders fought. Possibly it struck a sword guard, as re-created in the right photograph.

based on all of the available data, this is the most likely scenario for this musket ball.

A second musket ball was found in a conflict area where the same Forty-Second Regiment of Foot were trapped in an orchard a little later during the conflict and were being sniped at by handpicked Continental troops. As with the musket ball in figure 3.20, the ball shown in figure 3.21 is also 0.61 inches in calculated diameter, suggesting that it was fired by an American. It also has a channel impression, but the channel is very narrow and terminates in more of a "V" at the bottom of the channel. This channel is much too narrow to be associated with the most common agricultural equipment, but the possibility cannot be ruled out totally. The most likely candidate is a sharp-edged metal object. As can be seen in the photos, the lead is "peeled" back, indicating that the ball hit an object head-on rather than being sliced by an edged weapon before the battle or being accidentally struck by a hunting arrow many years or decades after the battle (this location of the park has been used for hunting for primarily white-tailed deer). One possibility is that the ball struck the edge of a Highlander sword. If it had hit a sword edge directly after being fired, the velocity would have most likely bisected the soft lead ball. The ball in figure 3.21 is deformed from impact on the opposite side from the channel, as shown in the center photo. This might indicate that it may have first hit the ground and bounced, thus reducing its velocity and force before hitting an edged weapon, as simulated in the far right photograph.

Although the analyses of these two musket balls are speculative, based on the shape and characteristics of the deformations, the conclusions are drawn from the most probable scenarios. Possibly future forensic techniques will be developed to more positively identify what musket balls actually hit.

Midair Collision

On very rare occasions, two bullets fired from opposing forces collide in midair. If the two balls hit nearly or exactly in the same tangential plane, they can

3.21. Another musket ball with a channel-like deep but narrow impression, recovered at Monmouth Battlefield in an area where Scottish Highlanders fought. Possibly it struck a sword blade, as re-created in the right photograph.

MUSKET BALL AND SMALL SHOT IDENTIFICATION

3.22. Two musket balls fused together by a midair collision. They were excavated at a 1645 English Civil War site in Hampshire, England. *(Photograph by Sam Wilson, director of the Basingstoke Common Metal Detector Survey.)*

fuse together (figure 3.22). This type of event can be simply coincidental but can also be an indication of very intense opposing fire: the more lead shot in the air, the greater the chance of two hitting each other.

The musket balls shown in figure 3.22 appear to be about the same size. The total weight of this artifact is 45.2 grams, such that each ball weighs approximately 22.6 grams, assuming no lead loss. Using the Sivilich Formula, the calculated original diameters of each would be 0.63 inches. The following is a description of the history of the site and the discovery of the musket balls:

> Basing House was a Royalist stronghold besieged by the Parliamentarians on and off from 1643 to 1645, culminating in the final storming and destruction of the House led by Oliver Cromwell himself on 14th October 1645. The metal detector survey took place in an area of common land to the south of the main House where Parliamentarian siege fortifications are thought to be located and sought to recover evidence of military activity in this area. It also looked to explore land closer to the defences of the House which might indicate storming action. The find in question came from an area reasonably close to the House so may be related to this type of action or from general skirmishing in the "no man's land" between the two forces as we know there were numerous sallies, etc. (Sam Wilson, personal communication, March 29, 2015)

Using Impact Impressions to Interpret a Site

Knowing the diameter of a musket ball, we can determine what general type of gun it was used with. Being able to identify what it hit puts it in context with the location found. Looking at groups of musket balls of the same general sizes and

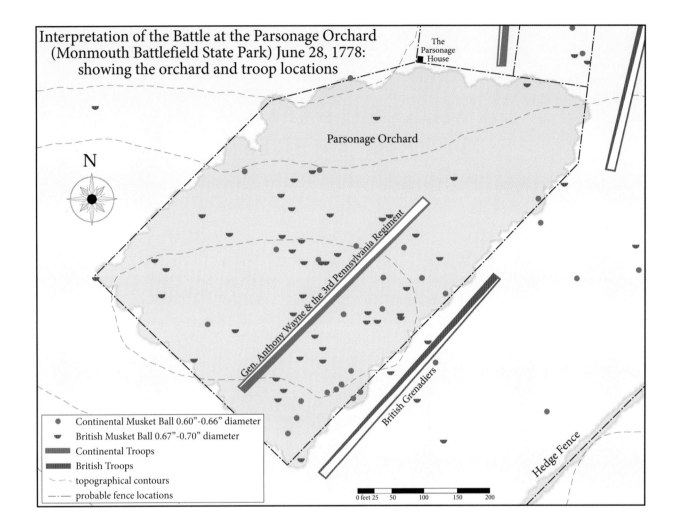

impact patterns can be used to identify site features such as tree, rocks, fence rails, and so forth. Combining all of this data yields a very accurate interpretation of the site.

To illustrate such a site interpretation, map 3.1 shows a section of Monmouth Battlefield State Park, New Jersey, where General Anthony Wayne and the Third Pennsylvania Regiment were forced into an orchard by British grenadiers. An examination of the group of musket balls that were fired from British Brown Bess muskets (diameters greater than 0.66 inches) and that were impacted and a second group of 0.60–0.66-inch-diameter musket balls from American (French) muskets that were dropped while reloading gave a clear approximation of the location of the orchard. Nearly all the musket balls of the British group appeared to have hit trees. The Americans had good cover in the orchard, while the British were in the open. Taking fire from the Continentals and from American artillery that had set up on an opposing hill, the British withdrew.

Map 3.1. Interpretation of the June 28, 1778, battle at the Parsonage Orchard (now in Monmouth Battlefield State Park) showing the orchard and troop locations (Continental troops in blue, British troops in red). *(Based on a map by Garry Wheeler Stone.)*

65

Chapter 4

Musket Balls with Fabric Impressions

Loading a lead ball wrapped in a small piece of greased fabric known as a "patch" creates a tight fit between ball and barrel. A patched ball captures more explosive force than a loose-fitting ball and has greater range and accuracy. This is particularly important for rifle balls, as it is the greased patch that engages the spiral grooves in the barrel. While patched musket balls may have been used when hunting, they were not commonly used in linear formation combat. Loading a patched ball could be too time-consuming in the heat of a battle.

It was widely known fairly early that reducing the windage between a musket ball and the walls of a musket barrel would increase distance and accuracy (Hogg 1981: 45). The better the gas seal, the greater the force of the expanding gas after the black powder is ignited. Because a residue of unburnt powder builds on the barrel walls with every shot, tight-fitting musket balls can be easily loaded with the first few shots, but as the gun becomes more fouled, loading the same size ball is increasingly difficult. Ramming becomes harder with each shot, and there is a potential of getting a ball stuck partway down the barrel. This situation can be dangerous because the gas has room to build excessive pressure, which can stress the barrel. This could lead to the barrel blowing up like a pipe bomb, causing injury or even death to the shooter.

Hunting muskets (often referred to fowling pieces) often came with a musket ball gang mold (figure 4.1) that had gradually decreasing-diameter ball cavities. Multiple cavities are used to cast a "gang" of musket balls or small shot. As described earlier, a smaller-diameter musket ball would be used for each consecutive shot. However, this would not be practical for military use. When facing a charging enemy during a battle, trying to select which size of musket ball to load next would be difficult, to say the least.

Musket balls with 0.05–0.10-inch windage were used for the military (Neumann 1967: 52). Typically, a musket ball can simply be dropped into a clean barrel with little or no ramming. The smoothbore was a more practical weapon for use in a battle line. When the enemy was advancing, the object was

4.1. Musket ball and buckshot molds as part of the 1655–75 Seneca village Dann site, now in the collections of the Rochester Museum and Science Center, Rochester, New York.

MUSKET BALLS WITH FABRIC IMPRESSIONS

to "get the lead out." Between rounds, the enemy advanced; the longer it takes to load, the quicker the enemy made gains. Aiming to kill was far less important than speed of firing, and wounding an enemy soldier tied up more personnel. In noncombat military (picket duty, on the march, and so forth) and even in civilian cases, there was an obvious advantage to keeping the musket ball tight against the powder charge. Having a loose musket ball could lead to problems such as having the ball roll out if the gun was inverted. To compensate for this, a musket ball could be wrapped in a fabric patch or wadded by ramming a fabric patch or an empty cartridge paper wrapper down the barrel on top of the musket ball. Both methods would keep the musket ball from moving in the barrel (Jim Stinson, personal communication, January 17, 2013). The patch method would also reduce the windage and yield a more accurate shot, but the patched musket ball would take longer to load.

Examples of wadded musket balls can be seen in figure 4.2. Artifact RC-6 (figure 4.2, left photograph) was found by Robert Campbell, a local metal detectorist, in a farm field in Burlington County, New Jersey. Although several other musket balls were found in the same field, there is not sufficient evidence to indicate whether these were associated with military activity. Clear fabric impressions can be seen in the two ramrod depressions. The thread count suggests that this was probably a linen wad. The ball was rammed hard enough to cause the thread pattern to become embedded in the lead. Artifact 232-11 (figure 4.2, right photograph) was excavated in Poland at the site of the 1759 Battle of Kunersdorf. This was one of the battles of the Seven Years' War and was Frederick the Great's largest defeat. Musket ball 232-11 was probably loaded and wadded just before the soldier went into battle. The three circular depressions indicate that this ball was tight fitting and required three hard strikes from a ramrod. As with artifact RC-6, the wadding left a very clear impression of the fabric weave.

4.2. Musket balls with fabric wadding marks in the ramrod indentations. The left musket ball was found by Bob Campbell on a farm in Burlington County, New Jersey. The right specimen was excavated at the site of the 1759 Battle of Kunersdorf, Poland. *(Left photograph by author, right photograph by Pawel Kobek and provided by Jakub Wrzosek of the National Heritage Board of Poland.)*

MUSKET BALL AND SMALL SHOT IDENTIFICATION

Nearly all lead balls used in early rifles were patched. A rifle has a series of grooves that are scribed in a helical pattern on the inside of a gun barrel. The purpose is to put a spin on the ball, creating a gyroscopic effect, which would stabilize the flight path to improve accuracy. Fouling also became a major problem, because the effective diameter of the barrel became reduced. To overcome this issue, balls were patched for use in rifles. The greased patch provided a good gas seal since the fabric was flexible, would fill the grooves, and acted as a gasket around the musket ball. The patch also acted as a wipe to help clean out fouling from previous shots. This became such an integral aspect of using rifles that many gun manufacturers would include a patch box built into the stock of the rifle to store fabric patches (Neumann 1967: 134).

The main advantage of a rifle over a smoothbore musket was its accuracy. In the hands of a common soldier, the maximum effective range (meaning the ball would kill or wound) was out to about 100–120 yards. In comparison, the maximum effective range of a rifle could be over 200 yards (Lawrence Babits, personal communication, October 1, 2013). However, the rifle was very slow to load. The ball had to be wrapped in a linen patch and forced down the barrel with a ramrod. For quicker loading, small wooden paddles with holes in them known as loading blocks (Neumann and Kravic 1989: 159) could have been used. A musket ball wrapped into a greased patch was forced into each hole. A ball in the loading block was placed over the muzzle of the rifle and pushed out of the board into the barrel. This technique is demonstrated quite accurately in the 1992 movie *Last of the Mohicans*. Rifles were made for accuracy, not speed. As such, they did not make a good firing line weapon, because the enemy could advance rapidly between loadings. Instead, rifles were good for sniping. The Continental army had groups such as Brigadier General Daniel Morgan's company of riflemen. There were many times when riflemen were used in battles for specific purposes. Morgan used militia riflemen at the Battle of Cowpens as the first line of battle to pick off Lieutenant Colonel Banastre Tarleton's advancing dragoons, targeting officers, and then to appear to retreat, drawing the British troops into a second line of volley fire from militia units. Despite heavy casualties, the British pressed on. Not having bayonets or sufficient time to reload, the American militia fled (Babits 1998: 87–99).

At the June 28, 1778, Battle of Monmouth, the Continental and British armies clashed in one of the largest battles of the Revolutionary War. In one segment of the battle, the second field battalion of the British Forty-Second Regiment of Foot, seasoned Scottish Highlanders, became pinned down in an orchard by American artillery fire. The orchard was in a swale that provided some cover for the Forty-Second from the case and canister shot. There they were additionally harassed by Continental marksmen sniping from behind a fence on the north side of the orchard. The Continentals had good cover from the fence line and could take their time loading their rifles and muskets, taking specific aim at a target. The archaeological surveys of this area by BRAVO recovered a large number of small-diameter rifle shot (less than 0.60 inches in diameter), confirming the sniping.

If a patched lead ball is rammed down a barrel with sufficient force, the black powder will be compressed into a concave, incompressible mass, and threads of a fabric patch can be pressed into the soft lead (Stinson, personal communication, January 17, 2013). The force of the explosion might also contribute to the fabric threads being pushed into the lead. This process leaves the warp and weft pattern of the patch material embedded in the ball. A very good example of this is shown in figure 4.3. This is a patched musket ball (0.78 inches in diameter) excavated at the site of the 1677 Battle of Landskrona, Sweden. The strong barrel band appears to have striations from rifling. Bo Knarrström suggests that it is from a rifled snaphaunce musket (Knarrström 2006: 32–35). The tight weave pattern impressed into the lead suggests that the patch was probably a finer grade of linen.

Patched rifle balls appear to be more common than originally thought. In 2008, members of BRAVO worked with archaeologists Adrian Mandzy and Stephen McBride in locating artifacts associated with the 1782 Battle of Blue Licks at Blue Licks Battlefield State Resort Park in Carlisle, Kentucky. On August 10, 1782, a force of British Loyalists, Native Americans, and some supporting British regulars clashed with American militiamen at Blue Licks. Daniel Boone was present, and his son Isaac was killed during the battle. Using metal detectors under the guidance of Mandzy and various members of the Kentucky Department of Parks, members of BRAVO helped locate the direction of fire and types of weapons used. Most of the lead balls were less than 0.60 inches in diameter, indicating that they were from rifles. This suggests that most if not all of the fighting was between militiamen and Native Americans. The British infantrymen evidently did not participate in the firefight at this section of the battle.

This was a very unusual survey in that it was in an area normally restricted to the public. Short's goldenrod, one of the rarest plants in the world, grows along the old buffalo trace that cuts through the park. Our excavations had to be monitored by Joyce Bender and Zeb Weese, both of the Kentucky State Nature Preserves Commission, to ensure that no damage was done to the endangered species. The entire project was coordinated by John Patrick Downs of the Kentucky Department of Parks. Everyone became very excited and interested when actual battle-related artifacts were found. I lent my detector to John Downs and gave him a crash course in its use. After digging up numerous trash items, John recovered the musket ball shown in figure 4.4.

Balls fired from smoothbore muskets can occasionally hook like golf balls. (As discussed earlier, the function of rifling is to have the projectile spin, creating a gyroscopic effect that causes the ball to travel on a more linear path.) Typically the impact area on a rifle ball is opposite the patch marks. Figure 4.4 shows a

MUSKET BALLS WITH FABRIC IMPRESSIONS

4.3. Musket ball with fabric patch impression and rifling marks excavated at the Landskrona Battlefield, Sweden. *(Photograph by Bo Knarrström, PhD.)*

MUSKET BALL
AND SMALL SHOT
IDENTIFICATION

Figure 4.4. Two views of a musket ball with fabric patch impression excavated at the Blue Licks Battlefield State Resort Park, Kentucky. *(Left photograph by Adrian Mandzy, PhD, right photograph by the author.)*

perfect example of a patched rifle ball that was fired and impacted. The patch weave is impressed nicely into the lead on the trailing side of the ball, and the leading side hit a target. A trace of the barrel band can be seen in the right photograph, indicating that this rifle ball fit tightly in the barrel. There is no visible ramrod indentation, since it would have been on the leading edge and would have been obliterated by the force of impact. This rifle ball followed a very true flight path.

Adrian Mandzy investigated the type of fabric impression that was left on the ball and reported, "According to Dr. Ivanna Prots, an art restoration specialist who deals with seventeenth and eighteenth century cloth, the material was hand woven, relatively coarse, and had a simple weave (Ivanna Prots, 2008, pers. comm.). As the material itself has disappeared, chemical analysis is no longer possible, but it was most likely jute or linen, with a small possibility that it could also be hemp" (Mandzy et al. 2008).

Figure 4.5 is an example of a musket ball found by Glen Gunther at a Revolutionary War camp site occupied by both American and British soldiers at different times in New Jersey. Although the site was military, there appears to be no evidence of any conflict. The size of this musket ball suggests that it was from a rifle. It has excellent definition of the fabric patch around the trailing edge. The leading edge is slightly flattened, which indicates that it was fired. It may have been fired during foraging or simply fired into the ground to discharge a weapon. It may also be from a local hunter and not necessarily associated with the war, but it does give us an excellent view of the fabric patch used.

4.5. Lightly impacted musket ball with fabric patch impression recovered from a Revolutionary War camp site in New Jersey. *(Photograph by Glen Gunther.)*

Of course, according to Murphy's Law, for every rule there must be an exception. The musket ball in figure 4.6 is just that exception: it has a faint fabric impression on the leading edge in the impact area.

The hemispherical shape of the musket ball indicates that it hit a hard target. It is from a large-bore musket, probably a Brown Bess, based on the calculated diameter (Sivilich Formula). It was excavated at Monmouth Battlefield in a skirmish area with other musket balls and a fragment of a Brown Bess trigger

MUSKET BALLS WITH
FABRIC IMPRESSIONS

4.6. Two views of the same impacted large-diameter ball with fabric patch impression in the impact area excavated at Monmouth Battlefield State Park, New Jersey. *(Photographs by Eric Sivilich.)*

guard. There are several possibilities as to what caused the fabric to be impressed on the impact side of the artifact:

1. The musket ball was from a smoothbore musket and had been patched to keep the ball from rolling out. Because it was fired from a smoothbore, the lack of spin caused the ball to rotate and hit an object patch first.
2. The musket ball was not patched and was fired from a smoothbore musket, after which it hit a human bone such as the sternum (breastbone) from the front or the scapula (shoulder blade) from the rear, and the weave is an impression from the soldier's clothing.
3. The musket ball was not patched and was fired from a smoothbore musket, after which it hit a soldier's accoutrement, such as a haversack or backpack, with a solid object in it such as a pewter plate. The weave is an impression from the fabric.
4. None of the above.

The first scenario is somewhat plausible, but if the musket ball hit a hard object such as a tree, one would think that the fabric would be deeply pressed into the musket ball, unless the velocity dropped. This artifact was found in an area of modern farmland in which modern nitrate-based synthetic fertilizers were used. Nitrates are very corrosive to most metals, especially copper and lead (Craig and Anderson 2002). Personal observations at numerous sites have shown that metal artifacts found in areas where there is no evidence of chemical fertilizing exhibit much less degradation. Therefore, the weave pattern on the musket

MUSKET BALL AND SMALL SHOT IDENTIFICATION

ball may have been much deeper at the time of impact and with time some has eroded away.

The second scenario is not very plausible, because the musket ball would have had to pass completely through the body after going through hard bone and would only then have come to rest in the field. However, this was in a conflict area, and other artifacts have been found at Monmouth that made contact with a human body (such as the musket ball with the possible tooth impression shown in chapter 3). Without an analysis for human DNA or blood presence, this could be difficult to prove.

The flat impression of the impact face and the linen-like weave make the third scenario seem more plausible. Haversacks and some backpacks were typically made from coarse-woven linen. Soldiers carried most of their possessions in them. The impact pattern is flat, suggesting that the ball hit a thick, solid flat object. Some likely candidates could be a pewter plate or a thick wooden object. Although the determination of what exactly happened is speculative, the facts are that the musket ball was fired, that it did hit something, and that there is a distinct fabric impression at the point of contact.

CHAPTER 5

Musket Balls Altered to Improve Lethality

When a circular bullet passes through human flesh and does not hit bone, it can leave a "clean" round wound that has the potential for being repaired, and the victim has a reasonable chance of surviving. Modifying the shape or characteristic of the projectile can make it more lethal. Modern hollow-point or "dumdum" bullets are illegal in the United States and many other countries because the tip mushrooms on impact, causing major trauma inside the body, thus significantly increasing potential lethality. The same is true for musket balls. Lethality, or perceived lethality, is the objective of altering them. There is also a psychological impact of increased fear for the potential target.

Halved and Quartered Musket Balls

A simple method for increasing the actual or perceived lethality of a musket ball is to cut it in half and to load both halves or cut it nearly in half (split shot) so that it will expand and separate into two projectiles after being fired. However, the changes in surface characteristics and center of gravity of each half would most likely affect accuracy and velocity/force of impact from the two new projectiles. As such, this type of shot would be effective only at close quarters. Figure 5.1 shows examples of musket balls intentionally cut in half. The left photograph shows both halves of a bisected musket ball. They were found by Reginald Pelham Bolton and William Louis Calver, members of the Field Exploration Committee for the New-York Historical Society. The artifacts were excavated at a "Revolutionary War site" in New York City (that is, the actual site is unknown) at the turn of the twentieth century (Calver and Bolton 1950: 72–79). The right photograph in figure 5.1 is a half musket ball found at Monmouth Battlefield in an area where the Continental troops rested from the extreme heat while American and British artillery bombarded each other. The lack of deformation and the location where it was found indicate that it had never been fired. The 0.72-inch

MUSKET BALL
AND SMALL SHOT
IDENTIFICATION

5.1. Halved musket balls: *left*, excavated by the Field Exploration Committee in New York City (*Collection of the New-York Historical Society, in artifact group 1947.283.A–W*); *right*, excavated at Monmouth Battlefield State Park, New Jersey.

diameter is based on the measured diameter of the ball fragment, which may not reflect the original ball size and (because of the possibility that lead is missing) cannot be verified. The weight is 17.4 grams, roughly half that of a 0.72-inch ball. Comparing the weight of the artifact and calculating its diameter and checking the results against the approximated diameter of the musket ball are methods of helping identify whether a musket ball is halved and thus has a significant weight loss, or whether it is flattened from impact. The weight of artifact 205-10-982 indicates that its calculated original diameter would have been 0.58 inches if it was simply a flattened musket ball. By looking at the artifact, one can see that it is not impacted and has an outside arc radius greater than 0.58 inches; therefore, it must have been cut.

Besides being halved, musket balls were also quartered for the same reasons. The musket ball in the left photograph of figure 5.2 is a superb specimen found by William Calver and Reginald Bolton somewhere around New York City. Calver and Bolton specialized in excavating camp sites, so it is likely that this musket ball came from one. It was obviously never fired. The musket balls in the center photograph in figure 5.2 were found approximately 650 feet apart at Monmouth Battlefield State Park. Artifact 234-1-426 has a small sheared section at the bottom, indicating that it had been cut in a fashion similar to the Calver and Bolton musket ball, being cut nearly but not all the way through, and separated after firing. Its original diameter, estimated to be 0.61 inches, is based on its measured radius. This is consistent for an American weapon. It weighs 5.68 grams, which is nearly one-quarter of the weight of a 0.61-inch musket ball. It was found along the retreat route of the Forty-Second Regiment of Foot and was most likely fired by the unit of handpicked Continental troops described by Joseph Plumb Martin in his memoirs (Martin 1988: 128–30).

Musket ball 90H14MP in figure 5.2 (center photograph) is 0.64 inches in estimated original diameter and weighs 12.05 grams. Although it appears to have been quartered, the weight suggests that this musket ball was halved and may have hit a flat target. It was found in 1990 during excavations at the Derrick Sutfin house. This house was standing at the time of the battle and is still standing today. This musket ball may have originally been cut in half but then hit the house, causing it to partially flatten and appear to be quartered.

Musket ball 211-7-179 (figure 5.2, right photograph) was also found at Monmouth Battlefield State Park. It weighs 4.50 grams and measures 0.63 inches in diameter. If the weight of the recovered piece is assumed to be one-fourth of the weight of the full ball, the calculated diameter would be 0.58 inches. Therefore, the actual original diameter would have been between these two values. Its slightly deformed shape and the difference between the measured and calculated diameters indicate that it was fired and soft impacted (that is, it probably hit dirt). What is most interesting is that it was found in the skirmish area where the Forty-Second Highland regiment first encountered and pursued an American battalion of soldiers selected from various regiments—many of whom were excellent marksmen. Musket ball 234-1-426 (figure 5.2, center photograph) was also fired by one of the Continentals at the Forty-Second, but later in the day as the British retreated. We may have recovered two musket balls from the same American unit that was in both firefights! Mutilating musket balls was not a common practice. Only a small number have been found at Monmouth as compared

MUSKET BALLS ALTERED TO IMPROVE LETHALITY

5.2. Quartered musket balls: *left*, excavated by the Field Exploration Committee in New York City (*Collection of the New-York Historical Society, in artifact group 1947.283.A–W*); *center* and *right*, excavated at Monmouth Battlefield State Park.

MUSKET BALL
AND SMALL SHOT
IDENTIFICATION

5.3. Two views of a quartered musket ball excavated at Fort Montgomery State Park, New York, in 1971. *(Photographs by Dana Linck.)*

to round musket balls. This may indicate that these two musket balls may have even been fired by the same American soldier.

These are examples of mutilated musket balls from the American Revolutionary War. The question becomes: which army was responsible for making these frightening projectiles? Figure 5.3 shows a quartered musket ball found at Fort Montgomery, New York. It is 0.69 inches in diameter, suggesting that it was from a Brown Bess musket. The fort was built by the Americans but captured by the British, who occupied it for a short time before destroying it. So which army is responsible? The poor patination and gray color suggest that it was made of an alloy including pewter, which may indicate that it was made for the Continental army (see chapter 8).

Musket Balls with Nails

Another way to make a musket ball more deadly was to drive a nail through it. This was probably as much a psychological weapon as it was a destructive projectile. Just thinking of such a projectile being fired at a human target is a frightful prospect, especially if you were the intended target. It would tear flesh and cause massive internal damage that would be nearly impossible for a surgeon to repair. As such, it has a very high probability of causing a fatal wound because of potential blood loss in a soldier even if it did not hit a vital organ. The nail would shift the center of gravity of the musket ball, causing it to wobble while in flight and significantly reducing its accuracy. With shoulder-to-shoulder linear warfare, accuracy was not necessarily critical. The existence of these altered projectiles is described as follows:

> Such were the peculiarly wicked missiles alluded to in General Lord Howe's complaint to Washington in September, 1776. Howe's communication reads:
>> My aid[e]-de-camp will present to you a ball cut and fixed to the end of a nail, taken from a number of the same kind, found in the encampment quitted by your troops on the 15th inst. I do not make any comment upon such unwarrantable and malicious practices, being well assured that the contrivance has not come to your knowledge.
>
> From his Headquarters on Harlem Heights, on September 23, 1776, General Washington replied to Lord Howe as follows:
>> Your aid[e]-de-camp delivered to me the ball you mention, which was the first of the kind I ever saw or heard of. You may depend [on it that] the contrivance is highly abhorred by me, and every measure shall be taken to prevent so wicked and infamous a practice being adopted in this Army. (Calver and Bolton 1950: 75–76)

Calver and Bolton reported that the musket ball shown in the left photograph of figure 5.4 was found in context with British military buttons in a refuse pit near a camp at 213th Street in Washington Heights, Manhattan. Ironically, they suggest that it appears to have been made by a British soldier. The practice of using a nail in a musket ball was apparently continued after the American Revolution and in other parts of the world. The center photograph in figure 5.4 shows a musket ball from the 1806 Battle of Pułtusk in Poland. It has a distinct square hole that was most likely from a wrought nail being driven through the ball.

The right and far right photographs of figure 5.4 are of a lead object recovered from Monmouth Battlefield State Park in a skirmish area. The hole is very square, indicating that the object was probably wrapped around a nail, as shown in the far right photograph. A hand-wrought reproduction horseshoe nail is shown, but the nail could have been in any style of the same thickness. It appears to match the description by General William Howe of being "cut and fixed to the end of a nail." It was found in the area crossed by the retreating British Forty-Second Regiment at the point where they crossed a fence line into an adjoining meadow. This information suggests that it would have been a musket ball made by wrapping lead (possibly from a musket ball that was intentionally flattened) around a nail and shaping it to the correct diameter for a specific musket. The unraveling indicates that it was fired. It is estimated to have originally been 0.62 inches in diameter, which would be consistent for use with a Charleville musket.

MUSKET BALL
AND SMALL SHOT
IDENTIFICATION

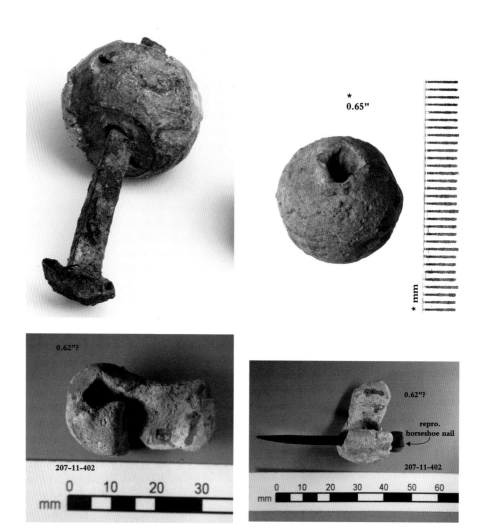

5.4. Musket balls with nails. *Top left:* Excavated by the Field Exploration Committee at Revolutionary War sites in New York City (*Collection of the New-York Historical Society, in artifact group 1947.283.A–W*). *Top right:* Excavated at the Pułtusk Battlefield, Poland. This musket ball has a distinct square hole that was most likely from a wrought nail (*photograph by Pawel Kobek and provided by Jakub Wrzosek of the National Heritage Board of Poland*). *Bottom left* and *bottom right:* A lead wrap recovered at Monmouth Battlefield State Park that appears to have been wrapped once around a nail. It is shown in the bottom right photograph with a reproduction horseshoe nail inserted.

The group of American skirmishers pursued the Forty-Second through the meadow, firing several more times at them. One American fired one of the quartered musket balls shown in figure 5.2 (artifact 234-1-426). As discussed in the next section, this group of Continentals seems to have had a propensity for using altered lead shot.

"Sluggs"—Cylindrical Musket Balls

Since metal detecting archaeological surveys began in Monmouth Battlefield State Park, twenty-five cylindrical-shaped lead projectiles have been found in areas associated with the 1778 battle. Many of them have surface facets indicating that they were made by hammering lead into cylinders. Several have concave depressions on one or both ends, which is a product of peening and suggests that they may have been made from musket balls. All cylinder shot that was not overly deformed from being impacted has measured diameters from 0.51 inches to 0.59 inches.

These were originally thought to be musket balls for smoothbore muskets that had been hammered to fit rifles, but no land and groove marks were observed on any of the specimens. Further investigations suggested that they were for smoothbore muskets. They were most likely wadded or patched with the cartridge paper to reduce windage. As such, the projectile would tumble in flight and could "keyhole" the intended victim. As with nail shot, this would rip through human flesh, causing irreparable damage: it would be nearly impossible to have a surgeon repair the wound, and internal damage would be substantial.

5.5. Examples of cylindrical shot excavated at Monmouth Battlefield State Park. They were all recovered from the same general area.

As with the other types of mutilated shot found at Monmouth Battlefield, more than half the cylinder shot was found associated with firefights between the Second Battalion of the Forty-Second Regiment of Foot and Americans from a battalion of "picked men" (Stone, Sivilich, and Lender 1996). Five were recovered in the skirmish area where the Forty-Second first encountered and pursued the Continentals. Ten were found in an orchard area occupied by the Second Battalion of the Forty-Second. While in the orchard, the Highlanders were sniped at by Continental marksmen—probably from the battalions of "picked men" that had retreated across this area. Figure 5.5 shows seven examples excavated in the orchard. The far right projectile in the photograph (artifact 234-2-1015) is shown close-up in figure 5.6 in comparison to artifact 211-7-170. They are nearly the same size and diameter (0.51 inches and 0.54 inches, respectively), and both appear to have been made from 0.63-inch musket balls, based on their weights. However, projectile 211-7-170 was found in the early skirmish area when the British were pursuing the Americans, and artifact 234-2-1015 was found in

5.6. Cylindrical shot of similar sizes and shapes excavated at Monmouth Battlefield State Park from two different areas of conflict between the Forty-Second Regiment of Foot and a hand-picked Continental unit.

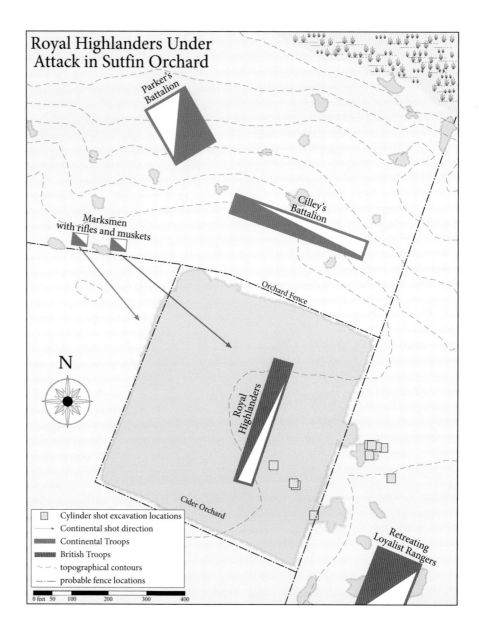

Map 5.1. Locations of cylindrical shot (shown as black rectangles) excavated from the Sutfin orchard site, from the attack on Royal Highlanders, now part of Monmouth Battlefield State Park. *(Based on a map by Garry Wheeler Stone.)*

the orchard area. They may have been fired by the same soldier, but without any distinguishing markings or chemical analysis of the lead impurities, this can only be postulated.

The cylinder shot recovered from the Sutfin orchard site at Monmouth appears to be confined to two distinct areas (map 5.1). The cylindrical musket balls were being fired by the American snipers at the Forty-Second Highlanders. Six of these projectiles appear to be concentrated in a specific spot, possibly the location of a Highland officer. The remaining four appear to be in a linear pattern, also possibly directed at a specific target. These patterns indicate that these cylindrical musket balls were being fired with a relatively high degree of accuracy and, along with the presence of quartered and halved shot and a musket ball made with a nail in it, suggest that these handpicked men were using shot designed to have a higher degree of fatality than spherical musket balls. These soldiers meant business.

Initially these hammered musket balls seemed unique to Monmouth, but similar ones were recovered by Historic Shipwrecks, Inc., on the pirate ship *Whydah*, which sank off the coast of Cape Cod in 1717. A black leather pouch containing twenty-three spherical musket balls and five cylinder shot was discovered, and these are now on display at the Whydah Pirate Museum in Provincetown, Massachusetts (Webster 1999). Ken Kinkor, curator of the museum at the time, allowed me to examine and measure all of the shot that was in the pouch. Table 5.1 shows the results for the cylinder shot.

The twenty-three spherical musket balls are all 0.63 inches in diameter. The conclusions are as follows:

1. Cylindrical lead shot was not unique to the Battle of Monmouth (1778) but had much older roots.
2. Cylindrical shot from the *Whydah* was made by hammering down 0.63-inch musket balls.
3. The shot in the *Whydah* pouch were for the same smoothbore musket and were not being used for a smaller-bore musket or rifle.

The left photograph in figure 5.7 shows two of the five slugs found in the same leather pouch as the round musket balls. The right photograph shows cylinder shot found in other sections of the ship.

The late Ken Kinkor's research on pirates uncovered a reference to our cylindrical shot in a reprint of the journal of Robert Drury (possibly a pirate), who was shipwrecked in 1703 with several other shipmates on Madagascar for fifteen years: "I made ready Deaan Afferre's and my own gun, and cast shot, or rather slugs, by making a hole in clay with a round stick to cast the lead in, and cutting it in pieces about half an inch long" (Drury and Rochon, 1969:163).

Table 5.1. Cylindrical shot excavated from the *Whydah*, 1717

Lead shot #	Weight (g)	Calculated diameter (in.)	Approx. measured diameter (in.)	Length (in.)
1	22.71	0.632	0.52	0.689
2	22.55	0.631	0.48	0.829
3	22.43	0.629	0.51	0.770
4	22.35	0.629	0.48	0.811
5	22.05	0.626	0.48	0.775
Average	22.42	0.629	0.49	0.775
Std. dev.	0.25	0.002	0.02	0.054

Note: Both ends of all the cylinder shot are convex.

MUSKET BALL
AND SMALL SHOT
IDENTIFICATION

5.7. Examples of the cylindrical shot recovered from the pirate ship *Whydah*, which sank in 1717. *(Left photograph by Ken Kinkor, right photograph by Christ Macort.)*

This account provides a name for our cylindrical projectiles—"slugs"—but were they legal to use? Ken Kinkor provided a pirate trial transcript of John Baptist Jedre in Boston. The transcript was printed by "T. Fleet, for S. Gerrish, at the lower end of Cornhill, 1726." The transcript included the following statement: "[T]he next Morning early they saw a Scooner, which they supposed at First to be an English Vessel, but when they came near they discovered she was a French Scooner which . . . they had traded with some Days before at Malegash . . . [W]hen they First espied her, they Cut up the English Fishing Leads, and beat them into Sluggs, divided among them the English Men's small Arms, and loaded them with the Sluggs." The fact that "sluggs" are specifically mentioned in the trial suggests that they may have been considered malicious or possibly illegal. This also states that they were beaten into shape. This additionally identifies the facets on the artifacts found at Monmouth and other sites.

This brings up a question: how commonly used were slugs? Once properly identified, slugs began to show up in other locations. The left photograph in figure 5.8 shows a musket ball hammered into a slug that was excavated at the Battle of Cooch's Bridge in Newark, Delaware. The battle, which took place on September 3, 1777, was the only American Revolutionary War battle in Delaware. The telltale flat spots on this artifact are clearly visible. The two cylindrical shot shown in the right photograph of figure 5.8 were found by Robert Campbell in Lumberton, New Jersey, in two different fields. Artifact RC-3 was found in the same sod farm field as the altered musket balls shown in figures 5.18 and 5.19, along with conventional musket balls. Although this general area had much activity during the American War for Independence, precisely what occurred at this specific site is not known.

Were these cylindrical shot found only in early American sites? I put a request out to my friends belonging to the Conflict Archaeology International

MUSKET BALLS
ALTERED TO
IMPROVE LETHALITY

5.8. Musket balls hammered into cylindrical shot: *left*, excavated at the site of the 1777 Battle of Cooch's Bridge, Delaware; *right*, found by Bob Campbell in two different locations in Lumberton, New Jersey.

Research Network (CAIRN), and the results were startling. Damian Shiels of Rubicon Heritage Services in Little Island, County Cork, Ireland, reported on a slug that was excavated at the 1691 Aughrim Battlefield in Ireland. It appears to have been hammered and is convex on both ends (figure 5.9). There is no indication that it was fired. It weighs 32.0 grams, which yields a calculated original diameter of 0.71 inches.

The Battle of Aughrim was a decisive battle of the Williamite War in Ireland. It was fought between the Jacobites and the forces of William III (or Williamites) on July 12, 1691, near the village of Aughrim in County Galway.

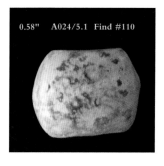

5.9. Cylindrical shot excavated at the site of the 1691 Battle of Aughrim in Ireland. *(Photograph by Damian Shiels.)*

> The bullet was found as part of a metal detector survey carried out on behalf of the National Roads Authority and Galway County Council as part of the N6 Galway to Ballinasloe road scheme, I analysed and reported on the bullets recovered, which relate to the 1691 Battle of Aughrim. They were found in a peripheral area of the battlefield, known as Lutrell's Pass, and the assemblage most likely represents part of the rout of Jacobite forces as they fled Kilcommodan Hill having been broken by the Williamites. I actually noted in the report that it is very similar to examples you got at Monmouth! It is a hammered cylindrical ball, as this can be seen at either end, and was a larger calibre hammered down, presumably to accommodate its use in a weapon with a smaller bore. I identified it as a probable slug—I discussed it with Glenn Foard as well who thinks it may be associated with mounted troops, looking to pack more punch in close combat, but this isn't confirmed. (Damian Shiels, personal communication, January 26, 2012)

MUSKET BALL
AND SMALL SHOT
IDENTIFICATION

Xavier-Rubio Campillo did his dissertation on the 1714 Battle of Talamanca, Spain, where he excavated five specimens of slugs. They appear to have been hammered and were crudely made (figure 5.10).

Two examples of cylinder shot or slugs have been recovered that were more carefully hammered (figure 5.11 on page 85). They appear to have been swaged or rolled to create a flat band. They could almost be confused with a ball having a very wide barrel band, except that there is no evidence of a ramrod mark or any deformation from being fired. Close examination does show evidence of hammering. The musket ball or slug on the left has several distinct markings on one end that appear to have been intentionally incised; however, their purpose is unknown.

The two musket balls in figure 5.11 are similar to those shown in figure 5.12 on page 85, which were found thousands of miles away at the 1806 Pułtusk Battlefield in Poland. The left musket ball shown in figure 5.12 either was not fired or else hit soft dirt. The right musket ball shown in figure 5.12 shows impact deformation and possibly some scratches from the musket barrel. The distinct barrel bands and the lack of rifling marks are again an indication that they were being fired from smoothbore muskets.

Not all cylinder shot is made from musket balls hammered into shape. Figure 5.13 on page 85 shows multiple views of two cylindrical shot that were found at the Piedras Marcadas Pueblo site near Albuquerque, New Mexico. In 1540 captain general Francisco Vásquez de Coronado led his army against the indigenous Pueblo people to establish a base of operations in what was called the "Tiguex province" (Schmader 2014). The small diameters of these shot are consistent with early matchlock muskets. The projectile in the left photograph has some light striations that may be from rifling, and (although it is difficult to see in the photograph [*top row, right*]) the base appears to be stippled from a powder burn.

These cylinder shot are different because they appear to be made from flat lead strips that were rolled and hammered. Figure 5.14 on page 85 is a close-up of the top (or tip) of the cylinder shot shown in the right photograph of figure 5.13

5.10. Five cylindrical shot excavated at the 1714 Talamanca Battlefield, Spain (*from Campillo 2008: 23–38*).

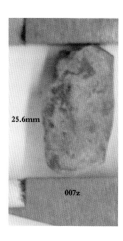

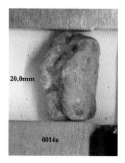

5.11. Two cylindrical shot found at an American camp site in New Jersey by Tim Reno. *(Photograph by Tim Reno.)*

5.12. Two musket balls that have been hammered or swaged into "sluggs." They were both excavated on Pułtusk Battlefield, Poland. *(Photograph by Pawel Kobek and provided by Jakub Wrzosek of the National Heritage Board of Poland.)*

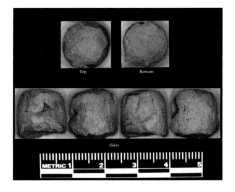
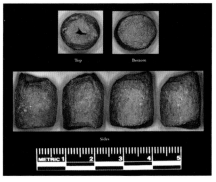

5.13. Cylinder shot from the Coronado expedition into New Mexico. *(Photographs courtesy of Matt Schmader, PhD, University of New Mexico.)*

5.14. Fabric impressions in the cylinder shot in the right photograph *(top row, left)* in figure 5.13. *(Photographs courtesy of Matt Schmader, PhD, University of New Mexico.)*

(top row, left). There is a distinct impression of a coarse-weave fabric that extends into the cavity of the projectile. The folds in the flat lead are visible, as is the overlap shown by the arrow.

Cylindrical shot or slugs were used throughout the world for a long period of time. They can be compared to modern dumdum bullets in that they were designed to inflict maximum damage on a human target. Unlike halved or quartered balls, their weight as a full musket ball would impact with substantial force, causing irregular entrance wounds, extensive internal damage, and a massive exit wound if the slug passed through a soldier's body. This design is for one purpose—to kill the intended target. As such, it would have also been a frightening psychological weapon.

Musket Balls with Small Cylindrical Cavities

Numerous musket balls have been recovered that have small holes that extend into the interior of the ball but do not penetrate completely through. Some are small pinhole cavities, probably made by air escaping from the molten lead during molding. Others appear to be larger "pinhole" shafts intentionally put into the lead ball, either during or after molding. The purpose of this deformation is not known, but several theories have been suggested:

- As air flowed over the hole, the projectile would whistle while in flight, making a psychological weapon (Ralph Phillips).
- The cavity could be stuffed with rancid meat or poison (Glenn Foard).
- This was an early form of a dumdum, or hollow point (author).
- The soldier was bored and decided to put a hole in his musket balls (author).
- It is a casting void (Glenn Foard).

An extreme example is shown in the left photograph of figure 5.15. It was excavated by the author at Monmouth Battlefield where a light woodlot once stood, the site of an intense battle that included volley fire and hand-to-hand combat between the Continental troops and British grenadiers. The musket ball is deformed from impact but has an unusual ovoid hole going partially into the ball. Its diameter of 0.69 inches suggests that it is from a musket such as the British Brown Bess, but because of the presence of numerous scattered musket balls in this area, identifying which side may have fired it is difficult. The shape of the cavity was determined by forcing clay into it. The clay cast shown in the right photograph has a tapered and slightly curved shape. This is possibly from the tip of a knife. It has a flat spot around the hole, suggesting that the hole was

MUSKET BALLS
ALTERED TO
IMPROVE LETHALITY

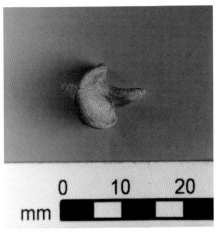

5.15. *Left:* Musket ball with cavity excavated at Monmouth Battlefield State Park. *Right:* Clay casting of the cavity.

made intentionally before being fired rather than being made by hitting a small pointed object.

Figure 5.16 shows three musket balls from two different occupation sites: Fort Montgomery and Valley Forge. Artifacts FtM10-906 (left photograph) and FtM10-936 (center photograph) were found approximately 145 feet apart at the Fort Montgomery site. One was at the top of a redoubt and the other on the steep slope below the redoubt. Neither shows evidence of being fired. Their diameters (0.58 inches and 0.55 inches) suggest that they were for rifles, indicating that they were from the Continental army. Artifact VF8-106 (right photograph) was found at the 1777 Continental army's Valley Forge encampment. The diameter suggests that it was for a Charleville musket. All three have flat areas that appear to have been made by removing the casting sprue with a knife. The mold seam is clearly visible on artifact FtM10-936 leading to the casting sprue. The holes appear to be from casting flaws: air bubbles during molding. The flat spot

5.16. Musket balls with possible casting flaw pinholes. The sprue in all of these appears to have been cut off with a knife.

where the casting sprue was is intriguing. No musket balls with carved flat spots without holes have been found at these sites or at any other. This seems to be more than just random chance. Glenn Foard of the University of Huddersfield in Queensgate, England, has found references to "spring shot" on the 1629 wreck of the ship *Batavia*. This shot was two musket balls joined by a wire, similar in form to artillery chain shot (Foard, personal communication, February 2–4, 2013). Spring shot was designed so that the balls would spin around the axis of the wire and inflict major trauma on the intended victim. Could the musket balls in figure 5.16 be spring shot (two musket balls joined by a length of wire) after the wire has corroded away?

Figure 5.17 shows three musket balls found in different locations. Artifact WTC-WF10S-002-25 was found on the remnants of a late eighteenth- to early nineteenth-century ship found at the base of the World Trade Center site. It has a distinct flat spot and an obvious small casting flaw. Artifact WP4-6-848 was found at Redoubt 4 at the U.S. Military Academy at West Point. It also has a casting void or bubble with a small flat spot. Artifact VF9-019 was found at Valley Forge. The casting flaw in the latter ball appears to be in the middle of a gouge made by an animal, probably a pig, which found and bit the artifact while looking for food. This area has not been exposed to artificial fertilizers or other corrosive materials. The heavy crazing of the surface of this lead musket ball may indicate that it was swallowed and exposed to gastric acids. These three balls also lack a sprue, and the left and center balls have distinctive flat spots.

Other Altered Musket Balls

I have had many discussions about the origins of small circular cavities with many colleagues belonging to CAIRN. The general consensus is that the holes

5.17. Musket balls with possible casting flaws.

5.18. Two views of round musket ball with 0.060" diameter wire passing through center of sprue and completely through the ball, found in Burlington County, New Jersey, by Bob Campbell. *(Photographs by Eric Sivilich.)*

are casting flaws. However, after examining the figure 5.16 photographs, Glenn Foard suggested that these musket balls may be spring shot and that the holes may be from wire. Figure 5.18 shows a musket ball that still has a remnant of an iron wire going through it. It was found in a farm field in Burlington County, New Jersey, by Robert Campbell, a metal detecting enthusiast. Although many other musket balls have been found at this site, there is no documentary evidence of military activity here. During the American Revolutionary War, New Jersey had a mixed population of Whigs and Tories. Numerous conflicts occurred between local groups that were not recorded. The number of lead shot found suggests that this may have been a short-term camp site. This musket ball has a diameter of 0.72 inches, which indicates a large-bore musket. It is possibly from an early Committee of Safety Brown Bess, but this is purely speculative. The musket ball is shown for informational purposes. The nonspherical shape is an indication that it was fired and hit a soft target. The site is a coastal plain region, and the soils are nearly pure sand with very small pebbles.

Figure 5.19 is a photograph of two musket balls that appear to be fused or molded together. This artifact was found by Robert Campbell in the same farm field in Burlington County, New Jersey, as the musket ball in figure 5.18. The perfect alignment of the mold seams indicates that these were molded as one double shot rather than being fused. The diameters of the balls and the grayish color and blistering suggest that this is a lead/pewter alloy, which would suggest an American origin. Similar types of double shot have been found at several seventeenth-century English Civil War sites (Foard 2012).

This style of ordnance was made for the same purposes as the cylinder shot: to tumble and cause massive trauma upon impact.

5.19. Double shot musket ball that was cast or fused together, found in Burlington County, New Jersey, by Bob Campbell. The mold seam aligns exactly on both balls.

MUSKET BALL AND SMALL SHOT IDENTIFICATION

Flint Wraps

Gunflints are knapped by removing flint blades from a larger rock known as a core. The blades are then shaped into rectangular gunflints (or spalls) by further knapping and pressure flaking. In use, the gunflint is put between the jaws of the cock and the jaw screw is tightened to hold the flint in place. The hard surfaces of the flint are smooth and not necessarily exactly parallel to the steel jaws, and the flint can rotate easily in the jaws or crack if the jaw screw is turned too tight. To overcome this problem, a buffer between the flint and jaws is required, as shown in figure 5.20. These "flint wraps" were typically made of lead or leather.

Leather decomposes easily; thus, leather flint wraps are usually not excavated at military sites. Figures 1.13–1.16 show a French musket lock excavated at Valley Forge that has a flint in the cock but no flint wrap; the leather wrap appears to have decomposed.

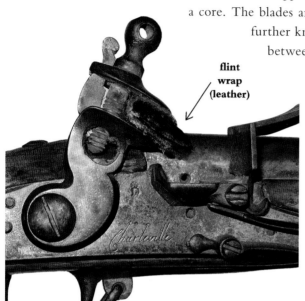

5.20. A Model 1768 French military musket cock with a leather flint wrap. (Photograph by Bill Ahearn.)

In contrast, lead flint wraps are found at military sites. Lead flint wraps can be manufactured from cutting sheet lead or flattening a musket ball. Figure 5.21 demonstrates how the flint fits into the wrap even if the flint is slightly larger than can be covered by the lead. This artifact was excavated at Fort Hunter, which was built during the French and Indian War on the Susquehanna River in Dauphin County, Pennsylvania. It was found as shown with the flint still in the wrap.

Figure 5.22 shows two gunflint wraps excavated at Monmouth Battlefield from two different segments of the battle. Both are nearly the same weight and appear to have been made from rifle balls. All three wraps appear to have smooth, rounded edges, suggesting that they were made from flattening spherical objects. If these wraps were made from sheet lead, the edges should be squared off, which would be evidence of being cut or sheared.

5.21. Flint with a lead wrap (artifact 36Da159) excavated at Fort Hunter, Pennsylvania. (Photograph courtesy of the State Museum of Pennsylvania, Pennsylvania Historical and Museum Commission.)

Many flint wraps can be differentiated from a flattened piece of lead because of a distinctive hole in the fold line. As can be seen in figure 5.20, the fold of the flint wrap presses against the jaw screw. After every firing, the flint is pushed farther toward the jaw screw, increasing the pressure and friction on the flint wrap. Figure 5.23 shows the front and back of a flint wrap excavated at the Washington Memorial Chapel site in Valley Forge, Pennsylvania. The left photograph shows that a hole may have worn through the lead by friction with the jaw screw. The flint would have been in contact with the steel jaw screw and could have chipped or cracked when the gun was fired. The square shape of this flint wrap suggests that it was probably made from sheet lead rather than from a flattened musket ball.

Figure 5.24 shows both sides of another flint wrap found at Valley Forge. The jaws of a musket cock were serrated to better grip the flint and wrap more securely. Not only is a hole worn from the jaw screw apparent, but also a small portion of the serrations was impressed in the lead, as seen in the left photograph. This must have been a high spot in the flint.

MUSKET BALLS ALTERED TO IMPROVE LETHALITY

5.22. These two flint wraps from Monmouth Battlefield were probably made from rifle balls. The calculated spherical diameters are shown, assuming no lead loss.

5.23. Two views of a worn flint wrap excavated at the Washington Memorial Chapel site in Valley Forge, Pennsylvania.

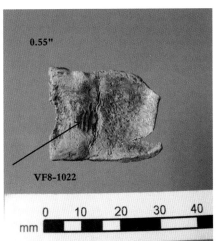

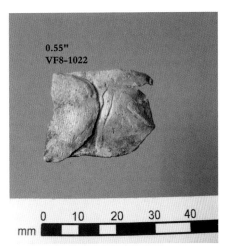

5.24. Two views of a worn flint wrap excavated at the Washington Memorial Chapel site in Valley Forge, showing the jaw serrations in the left photograph.

CHAPTER 6

CANISTER SHOT

WHEN IS A MUSKET BALL not a musket ball? When it is part of an artillery projectile!

On December 28, 1991, Ralph Phillips and I were doing a preliminary survey in the Sutfin orchard at Monmouth Battlefield State Park. Phillips found two musket balls with different diameters found fused together (see figure 6.1). We became very excited thinking we had found a rare artifact: a British musket ball and an American musket ball that had collided in midair and fused together! Artifacts resembling this had been found at American Civil War battle site, but we were not aware of any being found at Revolutionary War sites. Oddly, both musket balls had unusual, nearly cubic, shapes.

More metal detecting studies at the site turned up numerous unusually shaped musket balls as well as more fused shot (figure 6.2).

At that point, I was still not sure what these lead shot were, but I noticed that some were wedge shaped and others were more cubic. I asked my daughter, Michelle (then a teenager), to look at different arrangements to see whether they "fit" together. She placed the wedge-shaped ones next to each other until they made a circle. The cubic musket balls sat well in the center of the circle. The arrangement that she did is shown in figure 6.3. The rough diameter was 2.75 inches. That is approximately the diameter of shot for a three-pounder cannon (Caruana 1979: 14), but because lead balls are compressed, these were more likely used with a four-pounder. The size/description of an early cannon was that of the weight of the iron cannonball that it was designed to shoot. It had become evident that this was artillery canister shot.

Further confirmation of the use of canister shot came directly from historical document research. Major General William Alexander, Lord Stirling, stated in an August 15, 1778, letter to his friend William Henry Drayton concerning the British Second Battalion of the Forty-Second Regiment of Foot, "Their infantry appeared also in the rear of Sutfins; some of them advanced to the front of the orchard. These we drove back with grape and canister" (New Jersey Historical Society 1942: 174).

6.1. Fused musket balls excavated at Monmouth Battlefield State Park, New Jersey. *(Photograph by Michael Smith.)*

6.2. Three pairs of fused musket balls excavated at Monmouth Battlefield State Park. *(Photograph by Michael Smith.)*

CANISTER SHOT

What exactly is canister shot? There has been much interchangeability between the terms "canister shot," "case shot," and "grape shot." In 1779, *An Universal Military Dictionary* defined case shot as follows:

> Tin-case-shot, in artillery, is formed by putting a great quantity of small iron shot into a cylindrical tin-box, called a canister, that just fits the bore of the gun. Leaden bullets are sometimes observed, that whatever number of sizes of the shots are used, they must weigh, with their cases, nearly as much as the shot of the piece.
>
> Case-shot, formerly, consisted of all kinds of old iron, stones, musket-balls, nails, &c. and used as above. (Smith 1779: 141)

For the purposes of this book, canister shot is defined as a tin can filled with lead balls. Case shot is a tin can filled with iron balls. Grapeshot has a wooden base and spindle around which iron balls are stacked, put into a fabric bag, and wrapped with twine to hold it together. Mixed case shot is a tin can filled with lead musket balls and iron balls. All types were designed to be antipersonnel weapons used in artillery. Figure 6.4 shows examples of a canister shot tin case and a nearly complete canister shot projectile.

Cannonballs are not very effective as antipersonnel devices. They are long-distance projectiles designed primarily for damaging structures such as defensive walls and ships' hulls. The concept of using multiple small projectiles that would scatter in a wide fan when exiting the cannon muzzle was much more effective against advancing troops. In the sixteenth century, "hail shot" is reported as being used. "Tyn cases fill'd with Musquett Shott" were included in

6.3. Musket balls used as canister shot arranged by Michelle Sivilich in a reproduction tin can. They were all excavated at the Sutfin orchard site at Monmouth Battlefield State Park. *(Photograph by Michael Smith.)*

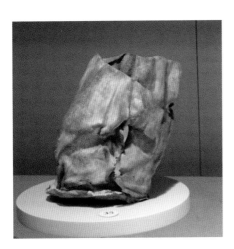 

6.4. Canister shot containers: *left,* tin container excavated at Fort Niagara, c. 1759–1812), on display at the Rochester Museum and Science Center (now in the collections of the Rochester Museum and Science Center, Rochester, New York); *right,* nearly complete canister with large lead shot, on display at the Muzeum Wojska Polskiego in Warsaw, Poland *(photograph by Adrian Mandzy, PhD).*

MUSKET BALL AND SMALL SHOT IDENTIFICATION

a 1635 inventory of stores at the Tower of London and in subsequent inventories up to 1725–26 (McConnell 1988: 319).

In 1800 Otto De Scheel in his *Memories d'artillery* described several kinds of canister shot:

> Of Cartouches, or Grape, Cannister and Other Compound Shot
> At the time these experiments were made, there was nothing well settled relative to cannon cartouches.—Two sorts were principally used.
> ... The first was composed of thirty-six iron or brass balls, arranged together on a wooden base, [a]round a pivot of the same metal, covered over with cloth, held together by a cord or wire, like the meshes of a net, the whole being tarred.
> The second was composed of musket ball and ammunition put together, without order and without attention to number, in a tin cannister [sic] placed on a wooden base, the diameter of which was regulated by the caliber of the piece for which it was intended.
> The first of these kinds of cartouch[e] took the name of grape shot, on account of its figure, and was made use of for twelve and sixteen pounders. The second was particularly used for eight and four pounders, and also occasionally for larger calibers. It is on this last that dependence is principally placed for throwing an enemy's line into confusion.
> As to the cannisters [sic] containing leaden ball, which were supposed to do great execution, and calculated for the calibers most in use, they were found to carry a still less distance. These balls were observed to adhere one to the other; sometimes they remained so, melted in the oddest forms, and they did their full effect in one lump; but instead of piercing through the platform, they only made an ordinary and weak contusion. (Williams 1984: 78–79)

6.5. Early nineteenth-century grapeshot for a four-pounder cannon, on display at the U.S. Military Academy West Point Museum, New York. *(Courtesy West Point Museum Collections.)*

This description matches the artifacts found at the Sutfin orchard site at Monmouth. De Scheel specifically stated that canister shot was used in "eight and four pounders." Joseph Plumb Martin, an Eighth Connecticut private who was engaged with the Forty-Second Regiment, wrote in his memoirs, "We had a four-pounder on the left of our pieces which kept a constant fire on the enemy" (Martin 1988: 129). Martin implies that this light field piece that fired a four-pound projectile was at the left end of the Continental gun line. General Alexander stated that grapeshot (figure 6.5) was being used by the same cannon.

As mentioned earlier, the terms "grapeshot" and "case shot" were often used interchangeably. The size of grapeshot specified for a four-pounder is two-ounce iron shot (Caruana 1979: 15). A large quantity of two-ounce grapeshot was recovered from the Sutfin orchard site.

Through battlefield archaeology and using the data gathered from both the excavated two-ounce grapeshot and canister shot, as well as descriptions in historical accounts, both the four-pounder position described by Martin and the Sutfin cider orchard where the Forty-Second Regiment of Foot was pinned down were located (map 6.1).

De Scheel also described the unusual shapes that the lead shot would take after being fired. Figure 6.6 shows specimens of canister shot excavated in the Sutfin orchard site at Monmouth Battlefield that have a variety of "unusual" shapes.

How can one tell the difference between canister shot and impacted musket balls? The key indication is the fact that many canister specimens have multiple facets and often have irregular shapes. These are caused by the musket balls compressing against each other. Usually the facets have very rounded edges. Impacted musket balls usually have only one or two facets, and the facets are typically flat with well-defined edges. The three main reasons for these deformations are described by Glenn Foard as formed "by compression and by melting in the barrel on firing . . . and by impact on hitting their target or bouncing

CANISTER SHOT

Map 6.1. Locations of canister shot and two-ounce grapeshot (case shot) excavated at Monmouth Battlefield State Park in an interpretation showing the hypothesized Sutfin orchard area, the location of the Forty-Second Regiment of Foot under fire, and the position of the four-pounder that fired the shot. *(Based on a map by Garry Wheeler Stone.)*

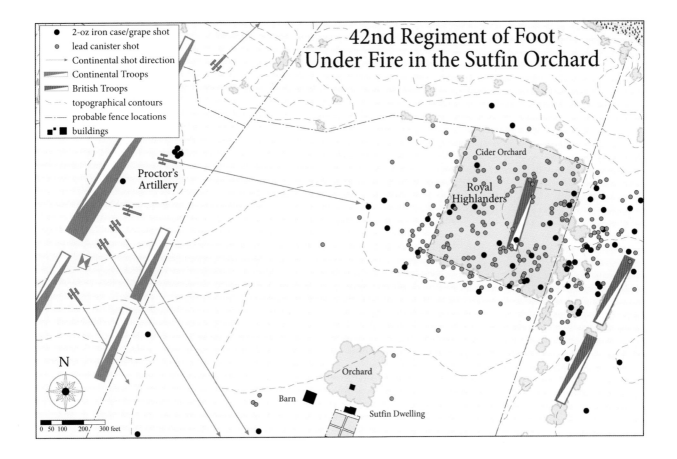

6.6. Canister shot excavated at Monmouth Battlefield State Park having different shapes.

on the ground" (Foard 2012: 89). Foard based his work on experimental firings of "hail-shot," an earlier form of canister shot using a wooden container.

Canister shot can be fired in two ways: point-blank range at zero elevation or ricochet firing (Adkin 2001: 269–70; Caruana 1990: 15). The first was used for very short distances. Depending on the thickness of the tin sheet used, the canister may or may not split open when exiting the muzzle of the cannon. In either case, De Scheel stated, "they did their full effect in one lump" (Williams 1984: 79). The second method of firing was to aim the projectile at a low angle intending to hit the ground in front of the enemy so that the shot would ricochet and spread into the enemy ranks at a wider angle. Analysis of the archaeological evidence at Monmouth Battlefield indicates that the later technique was being used (Sivilich and Sivilich 2010) in one section of the battle at the Parsonage Farm. A very good example is that of a Continental artillery position, under the command of General Nathaniel Greene, atop a hill on the Combs farm and firing iron case shot and lead canister across a ravine and onto the top of an opposing hill on the Parsonage Farm. The locations where lead shot and iron shot (map 6.2) were excavated are shown on a section of the Michel Capitaine du Chesnoy map (Chesnoy was the cartographer for the marquis de Lafayette). Note that the troop positions shown are earlier in the battle. The battle at the Parsonage Farm was fought later in the day and was between General Anthony Wayne's Pennsylvania regiments and the First Battalion of British grenadiers. With an estimated distance of more than five hundred yards to the target, the canister had to be fired with the cannon elevated and would have ricocheted upon impact and spread out for maximum effect. Between initial rapid acceleration and the later impact with the ground, the round musket balls would have been subjected to a great deal of compressive force and heat from the sudden decrease in kinetic energy. During the filming of the *Battlefield Detectives* television series episode "The Battle of Monmouth" for the History Channel, Glen Gunther of BRAVO excavated artifact 207-5-037 on camera (figure 6.7).

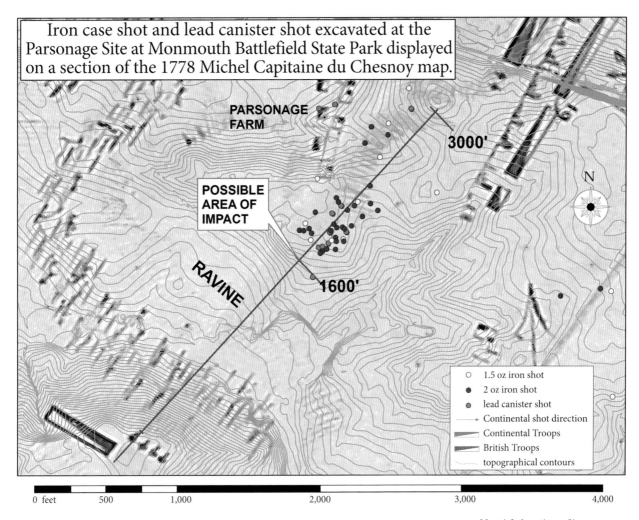

Map 6.2. Locations of iron case shot and lead canister shot excavated at the Parsonage site at Monmouth Battlefield State Park, displayed on a section of the Michel Capitaine du Chesnoy map in the Collections of the New-York Historical Society, reproduced by permission. *(Based on a map by the author.)*

6.7. Shot from the bottom corner of the tin canister excavated at Monmouth Battlefield State Park. The right photograph shows the original packing arrangement using reproduction musket balls of the same diameter.

MUSKET BALL AND SMALL SHOT IDENTIFICATION

One facet of this artifact was flat, one facet was slightly convex, and the three remaining facets were concave indentations. It was immediately identified by Garry Stone and me as an excellent specimen of canister shot. Its unique shape indicated that it was in the bottom outer row of the can, positioned on the wooden base and against the tin wall. It is shown in figure 6.7 as found (left photograph) and with three reproduction 0.69-inch musket balls placed on the concave facets (right photograph). The fit was perfect. That compressive forces shaped this artifact is obvious. However, as a simple matter of physics, two spheres of lead having the same hardness will create flat spots on both spheres at the interstices. Therefore, this suggests that the three lead balls that made concave indentations in artifact 207-5-037 had to have greater hardness values. How can this be possible?

As described in chapter 8, some musket balls were made with a tin alloy. This is harder than pure lead. As shown in figure 6.8, 100 percent lead has a Brinnel Hardness Number (BHN) of 7.0. A 90 percent lead/10 percent tin alloy has a BHN of 9.4 (Thompson 1930: 1093).

Because a lead/tin alloy is harder than pure lead, the compressive forces from such an alloy ball would create concave depressions in the lead ball. Twenty-two "lead/pewter" canister shot were also found at this site (figure 6.9). The Americans thus appear to have made canister shot from their stores of available musket balls rather than casting lead shot specifically for artillery projectiles.

Lead canister shot can be very diagnostic in analyzing battlefields. They are proof positive of light field artillery being used as antipersonnel weapons. The shot can be recognized by the specific deformed shapes caused by compression. Experimental artillery firings of seventeenth-century "hail-shot" were conducted by Glenn Foard at Warwick, England (Foard 2012: 89). Much of the lead shot was retrieved and examined. Foard concluded from these tests that

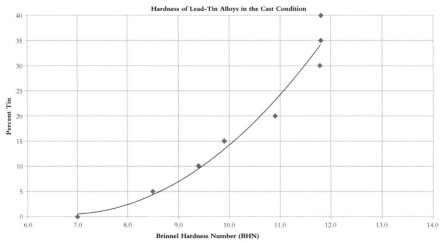

6.8. Lead-alloy hardness scale *(data from Thompson 1930: 10).* The greater the BHN, the harder the alloy.

the intensity of compression seen on hail-shot has been classified into four broad classes:

(1) round: showing no facets (in which case it is usually indistinguishable from small arms fire)
(2) faceted: showing small flat areas, usually widely spaced around the bullet, but with a mainly round surface . . .
(3) polygonal: where most of the surface comprises facets that join [and]
(4) flat: fully faceted but highly compressed in one direction.

Four examples of lead canister shot excavated at the 1759 Kunersdorf Battlefield in Poland have faceted (figure 6.10, left and right-center photographs), polygonal (figure 6.10, left-center photograph), and flattened shapes (figure 6.10, right photograph). The large calculated diameters of these artifacts are consistent with the Polish canister shown in the right photograph of figure 6.4.

CANISTER SHOT

6.9. Aerial photograph of the Sutfin orchard site (bordered by green box), Monmouth Battlefield State Park, showing locations of lead (blue dots) and lead-alloy (yellow dots) canister shot. Fused canisters are indicated by the number 2 in a circle.

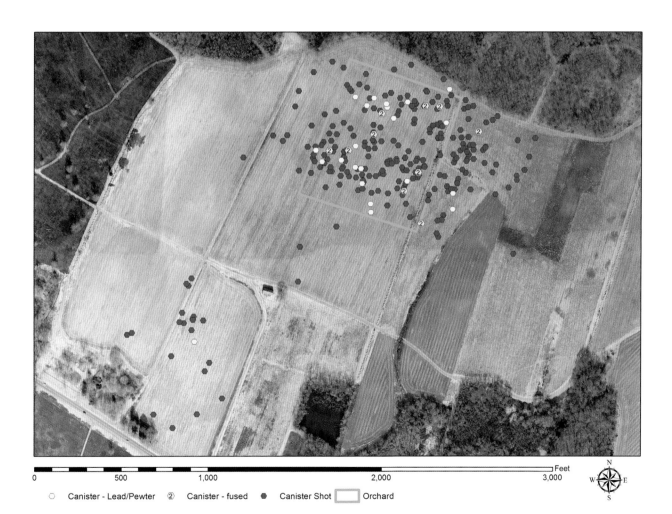

MUSKET BALL
AND SMALL SHOT
IDENTIFICATION

The canister shot in the right-center photograph in figure 6.10 also exhibits an unusual crosshatched impression such as a very coarse fabric weave or wood grain.

One must be very careful in identifying canister shot—not all faceted lead balls excavated at a military conflict site are necessarily canister. The concept of using round lead balls in an exploding projectile was invented by Henry Shrapnel, a British lieutenant, serving in the Royal Artillery in the mid-1780s (McConnell 1988: 324). It was also used extensively during the American Civil War. Before World War I, lead balls were hardened using antimony to improve target penetration. A July 31, 1915, article in the *Boston Evening Transcript* provides a description of shrapnel ball manufacturing. The following is an excerpt:

Shrapnel by the Ton [From Machinery]

The most deadly and effective parts of a shrapnel are the lead bullets which are held in the shell. When the timing fuse explodes the powder in the base of the shell, the nose is blown off and the bullets are thrown out in a cone shape. The range covered by these bullets in the 18-pound shrapnel shell is about 250 square yards. The lead bullets, which are in most shrapnel are ½ inch in diameter, are made from several different compositions, but chiefly consist of 87½ parts lead and 12½ parts antimony. The number of bullets carried in the shrapnel shells of the different governments varies. They are 252 in the American 15-pound shell and 235 or 236 in the British 15-pound shell. The bullets used by the U. S. Government have six flattened sides, to facilitate packing, whereas those used by foreign governments are spherical.

6.10. Four canister shot excavated at the site of the August 12, 1759, Battle of Kunersdorf, Poland, showing the different shapes attributable to compressive forces. Artifact 457-11a appears to have a wood-grain or fabric impression. *(Photographs by Pawel Kobek and provided by Jakub Wrzosek of the National Heritage Board of Poland.)*

Excavated antimony-hardened shrapnel balls do not have the same whitish patina as lead musket balls. Depending on the soils, they usually have little or no patination. Hardened balls have been recovered at several sites. For example, at Gateway National Park in Sandy Hook, New Jersey, BRAVO members assisted then–National Park Service archaeologist Dana Linck in excavating an area

 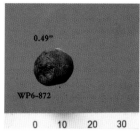 

CANISTER SHOT

6.11. Three examples of modern shrapnel balls excavated near Redoubt 4 at the U.S. Military Academy, West Point, New York, among eighteenth-century military artifacts.

designated for a new parking lot. This area had been an artillery ordinance testing area from 1875 to the Korean War. Numerous small gray lead balls with multiple facets were discovered. As another example, when I worked with Adrian Mandzy at the 1649 Battle of Zboriv in Ukraine, many small (approximately 0.50 inches in diameter) gray lead balls were found among white-patinaed musket balls. A previously unknown World War I battle was discovered on the same site as the seventeenth-century battle. Cast iron shell fragments and early twentieth-century brass rifle casings were recovered, confirming the World War I battle. In yet another instance, while I was working with Dana Linck at West Point near Redoubt 4, six specimens of small-diameter gray lead alloy balls were found in the eighteenth-century site. Several of these artifacts had six flat facets, as described in the *Boston Evening Transcript* article. The U.S. Military Academy at West Point, New York, is also an artillery school, and at some point the area was used for live fire. Figure 6.11 shows three examples of this hardened lead shrapnel. The distinct size, color, and shape of modern shrapnel should not be confused with its earlier counterparts. This can lead to interpretation errors if misdiagnosed.

CHAPTER 7

Chewed Musket Balls

There are many photographs on the Internet of musket balls that are so deformed from teeth impressions that they have been described as having been from field hospitals where the patients were in so much pain they were given musket balls to chew on. The balls are usually further described as being mashed so severely that the procedure on the patients must have been amputations. There is an example of "biting the bullet" on a dentistry products website referring to a musket ball that was on display at the National Museum of Dentistry (Levine 2012). Nearly all of these identifications are wrong: humans cannot bite into a pure lead musket ball hard enough to make very deep impressions. Now we will separate fact from fiction. Who or what made these impressions? The culprit of many misidentified chewed musket balls: swine.

Swine-Chewed Musket Balls

Swine have very powerful mandibles and very strong teeth. They are one of the few species that can crush, eat, and digest bone, including human bones. Why would pigs chew on musket balls? Pigs use their snout to root for food such as acorns, nuts, tubers, and other edibles that fall on or are buried in the ground. It could be days to decades after a military event occurred that either domestic swine or wild boars came through the area looking for food and could pick up and unintentionally chew a musket ball instead. Many conflict areas were farms that continued to be farmed long after a battle took place. Camp sites or engagements in remote areas were also subject to wild boars roaming for food. The southern United States today has a severe problem with wild boars being a threat to rural populated areas.

Zooarchaeologist Henry Miller, PhD, has identified dentition marks in many different musket balls. Attached as appendix B is an analysis he completed on identifying teeth marks from a variety of animals and humans in musket balls. Figure 7.1 shows examples of musket balls that have been severely crushed by swine,

CHEWED MUSKET BALLS

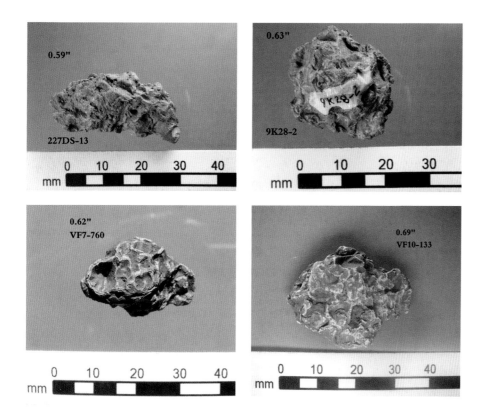

7.1. Severely mashed, swine-chewed musket balls: *Top left* and *top right*, found at the Point of Woods at Monmouth Battlefield, New Jersey; *bottom left* and *bottom-right*, excavated at the Washington Memorial Chapel site, Valley Forge, Pennsylvania. Calculated diameters shown in the photographs are based on the artifact weights but do not account for the possibility of lead loss.

using their molars to mash the lead balls like chewing gum. These were not chewed by soldiers who were having a limb amputated. These artifacts happened to be in an area of potential food sources where pigs or boars were rooting sometime after the musket balls were originally deposited. Swine-chewed musket balls have been recovered at numerous military sites. As one example, thirty-three animal chewed musket balls were found scattered across Monmouth Battlefield.

These severely crushed musket balls are relatively easy to identify as being swine-chewed, because of the deep molar impressions; however, swine will also chew objects using their incisors. Incisor gouges in musket balls can appear to be human-made, so one has to look carefully at the width and especially the depth of the indentations (for examples, see figure 7.2). The ball in the left photograph might very easily be mistaken as having been chewed by a human, but the depth and size of the bite marks are too deep and large to be human.

Some swine-chewed musket balls can be more difficult to identify because the definition of the edges of the teeth indentations are not very sharp. The three musket balls shown in figure 7.3 are examples of this type. They were excavated at Newark, Delaware, at the site of the September 1777 Battle of Cooch's Bridge, an area that was and still is farmland. This in itself is a good indication that it was a potential feeding area for domestic pigs at some point in time. Some musket

MUSKET BALL AND SMALL SHOT IDENTIFICATION

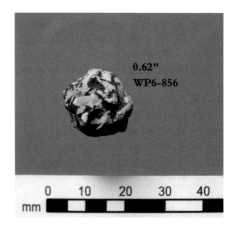 

7.2. Swine-chewed musket balls with incisor depressions: artifact WP6-856 (*left*) was found near Redoubt 4, U.S. Military Academy, West Point, New York; artifact VF10-203A (*right*) was excavated at the Washington Memorial Chapel site, Valley Forge.

7.3. Three swine-chewed musket balls with shallow dentition marks excavated at the site of the American Revolutionary War Battle of Cooch's Bridge in Newark, Delaware.

balls were very likely chewed by pigs and spit out, while others were swallowed and passed through the digestive system. Some chewed musket balls, like those shown in figure 7.3, may have been ingested and partially dissolved by gastric acids. This would obviously cause a loss in lead, giving a smaller-than-original calculated diameter.

Lead loss can give a false value if diameter is calculated from ball weight. Figure 7.4 shows two photographs of the same musket ball. It shows no signs of being impacted but definitely has a part missing, bitten completely off by a pig. This site was going to be destroyed by building construction, but because of the archaeological work done by myself and the Deep Search Metal Detecting Club, it was saved and is now part of the Monmouth Battlefield State Park.

Heavily chewed musket balls (figure 7.5) were excavated at the Smith's St. Leonard site, an early eighteenth-century plantation at Jefferson Patterson

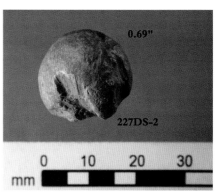

7.4. Swine-chewed musket ball with lead loss. Found by the author at the Point of Woods, 1778 Battle of Monmouth.

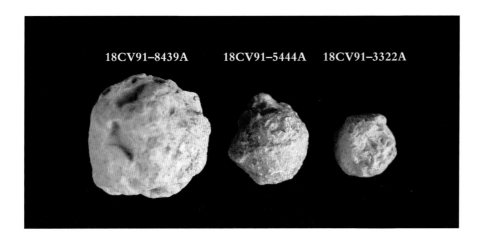

7.5. Swine-chewed musket ball from the Smith's St. Leonard plantation site in Maryland. *(Image provided by Edward Chaney, deputy director, MAC Lab; courtesy of the Jefferson Patterson Park and Museum.)*

Park and Museum in Maryland. The markings were identified as being from a swine.

In figure 7.6 we have our smoking gun . . . or musket! A pig tooth that was split in half was also recovered at the Smith's St. Leonard site. Embedded deeply into the crown of the tooth is a fragment of white metal that X-rays and visual examination determined was most likely lead.

Large-Rodent-Chewed Musket Balls

Rodent-chewed musket balls are another common type. Larger rodents such as rats and squirrels use their front incisors to gnaw objects. Rats will gnaw on many different materials. Squirrels have also been known to specifically eat lead, such as lead flashing around older vent pipes on roofs (Thomas 2011). Figure 7.7

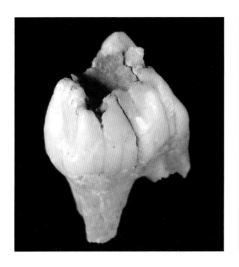 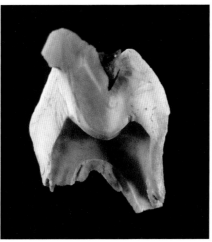

7.6. Swine molar fragment from the Smith's St. Leonard plantation site with lead embedded in the crown. *(Image courtesy of the Jefferson Patterson Park and Museum.)*

MUSKET BALL
AND SMALL SHOT
IDENTIFICATION

shows damage done to a section of garden hose by a rat. The parallel grooves of the front incisor bite marks are clearly shown.

Similar striations can be seen in figure 7.8, showing a lead gaming dice made from a musket ball. It was excavated at the Washington Memorial Chapel site in Valley Forge, Pennsylvania. The artifact does not have the customary dots but instead has Roman numerals. The "I" has has been obliterated by a rodent and the "II" partially chewed. Whether this was chewed by a rat or a squirrel is unknown, since their gnaw marks are very similar.

7.7. Rat damage to a section of garden hose showing the distinct striations of the incisors.

Why would a rodent chew on lead objects such as this dice? The most likely reason is that lead has a sweet, cool taste. Lead patina is often referred to as "sweet lead." However, there may be other reasons.

Figure 7.9 shows multiple views of the same musket ball that was chewed by a large rodent. Found at the Parsonage Farm site at Monmouth Battlefield, it was fired and is deformed from impact. After the battle, contemporary historical documents indicate that the Parsonage Farm was used as a field hospital. So the question becomes: what did this musket ball hit? If the ball hit a human target, it may have had blood on it. If it was embedded in a limb that was amputated and discarded, it could have easily attracted rats.

7.8. Rodent damage to two facets of a lead die excavated at the Washington Memorial Chapel site, Valley Forge.

7.9. Three views of the same impacted musket ball chewed by a large rodent; it was excavated at a place that was both an engagement location and a field hospital during the Battle of Monmouth.

7.10. Musket balls chewed by small rodents, such as mice: *left*, found at the Point of Woods at Monmouth Battlefield; *right*, found by Bob Hall at an American Revolutionary War encampment site in New Jersey.

SMALL-RODENT-CHEWED MUSKET BALLS

Small rodents, such as field mice, will also gnaw objects that interest them either because of their taste or because they are encountered while burrowing. Figure 7.10 shows musket balls recovered from two different locations that were chewed by small rodents, probably mice. The dentition patterns are very similar, radiating inward to a central point. The animal's jaw opening is smaller than the diameter of the musket ball, so the rodent can only nip and scrape at the ball. Coincidentally, both musket balls shown in figure 7.10 are chewed at the sprue.

DEER-CHEWED MUSKET BALLS

Pigs and rodents are not the only animals that chew on musket balls. Other animals looking for food, such as acorns, can accidentally pick up a musket ball and chew on it. The musket ball shown in the left photograph of figure 7.11 was excavated at Fort Montgomery, New York, where BRAVO has conducted several archaeological surveys. The last survey was with Dana Linck, a professional archaeologist who worked on the site in the 1960s. The site has several large oak trees, and large quantities of acorns were noted during excavations. This would attract various animals, including the North American white-tailed deer.

Linck has been analyzing the musket balls found at the site from earlier excavations as well as those found by BRAVO. He observed that musket ball FtM1971-210 (figure 7.11, left photograph) had unusual curved dentition impressions unlike the marks described earlier. Suspecting that the marks might be from a deer, he conducted an experiment using a deer jaw that had a complete row of teeth. He impressed the teeth into several soft clay balls (figure 7.11, right

MUSKET BALL AND SMALL SHOT IDENTIFICATION

7.11. Left: A musket ball excavated at Fort Montgomery, New York, with distinctive curved bite marks. *Right:* The marks were replicated in clay balls by Dana Linck using a white-tailed deer jaw with teeth. *(Photographs by Dana Linck.)*

photograph). He successfully duplicated the horseshoe-shaped markings and concluded that artifact FtM1971-210 was possibly chewed by a deer.

Human-Chewed Musket Balls

Did soldiers, or anyone else, ever "bite the bullet"—or is this just a myth? Many people immediately think of field surgery and amputations when they think of that phrase. There are actually numerous references to people chewing on and biting musket balls and for various reasons. A few examples follow.

Jeptha Root Simms recorded accounts from Revolutionary War soldiers in his research published as *The Frontiersmen of New York*. This is an excellent primary source account of biting on a musket ball to endure pain.

> Near West Point he saw a sergeant, a corporal, and two privates stripped and flogged one cold morning, each receiving one hundred lashes upon his bare back. These soldiers belonged to Gen. Glover's brigade, and left the army with several others to go home, insisting that their term of enlistment had expired. Their officers declared the time was not up, and the men were overtaken and brought back as deserters. The four mentioned were tried by court martial and punished; the two privates being the two youngest boys in their regiment. Pease, the sergeant, who was in Col. Shepard's regiment, had been a brave soldier, having served

under the daring Montgomery at Quebec. Much sympathy was felt among their fellow soldiers for these sufferers, particularly for the boys. The latter did not utter one word of complaint; *but each taking a leaden bullet in his mouth, bit upon it as the punishment was inflicted.* (Simms 1882: 590 [emphasis added])

Lieutenant John Waller of the British Marines wrote his brother soon after the Battle of Bunker Hill:

We had of our corps one major, 2 captains, and 3 lieutenants killed; 4 captains, and 3 lieutenants wounded: 2 serjeants, and 21 rank and file killed; and 3 serjeants and 79 privates wounded: and I suppose, upon the whole, we lost, killed and wounded, from 800 to 1000 men. We killed a number of the rebels, but the cover they fought under made their loss less considerable than it would otherwise have been. The army is in great spirits, and full of rage and ferocity at the rebellious rascals, who both poisoned and chewed the musket balls, in order to make them the more fatal. Many officers have died of their wounds, and others very ill: 'tis astonishing what a number of officers were hit on this occasion; but the officers were particularly aimed at. (Bell 2012)

Bell's account talks about chewing musket balls to improve their level of lethality. This is difficult to ascertain from an excavated musket ball that may have been a soft hit, because the chewing creates a certain level of deformation.

Thomas Mellen, a soldier at the August 16, 1777, Battle of Walloomsac, New York, wrote, "I soon started for a brook I saw a few rods behind, for I had drank nothing all day, and should have died of thirst if I had not chewed a bullet all that time" (Stark and Stark 1860: 67). This reference is very interesting and applicable to the Battle of Monmouth, which took place on June 28, 1778, one of the hottest days recorded that year. Lieutenant General Henry Clinton stated in a letter to his sister, "But the thermometer at 96—when people fell dead in the street, and even in their houses—what could be done at midday in a hot pine barren, loaded with everything that [the] poor soldier carries? It breaks my heart that I was obliged under those cruel circumstances to attempt it" (Clinton 1954: 94n15).

Many men died of heat fatigue that day. John U. Rees, who has conducted extensive research on the Battle of Monmouth, provided the following quotation from Andrew Bell, General Clinton's secretary: "[A]bout 200 killed and wounded—the weather destroyed more than the Action" (Bell 1778).

The quotation from Thomas Mellen indicates that chewing on a musket ball helps promote salivation. The following examples indicate that the back molars were used to bite down on a musket ball to help bear pain; however, front incisors are more likely to have been used to help promote salivation. Incisors cannot withstand as much bearing force as the more robust molars. Incisors are sharp and designed for cutting and tearing; molars are designed for crushing.

To determine how much deformation an adult male human can inflict on a lead musket ball by biting on it, several experiments were conducted by chewing on lead musket balls. (It should be noted that the first sign of lead poisoning is dementia, which might explain what drove me to actually write this book!) Zooarchaeologist Henry Miller and I chewed on reproduction musket balls that were cast from "pure" lead (Miller 2004). The results are shown in figure 7.12. The results were unexpected; we were not able to mash a musket ball into a flattened piece of lead but only dented the surface. We quickly learned that lead is not as soft as expected. A metal sphere is a difficult object to compress, since the atomic structure does not easily distribute laterally.

7.12. Reproduction musket balls chewed by the author and Henry Miller using incisors, canines, and molars.

Thanks to my daughter, Michelle (who has a PhD in military archaeology), I learned of XRF: portable X-ray fluorescence analyzers. This equipment can perform a nondestructive, qualitative metal alloy analysis to determine the metal composition. She alerted me that a company would be demonstrating their XRF equipment at the 2009 Society for Historical Archaeology conference in Toronto, Ontario, and a few samples could be tested at no charge; I took several musket balls with me to the conference. Thermo Fisher Scientific had a booth to demonstrate a Nitron XL3p analyzer and tested both the reproduction balls and actual musket balls that I had brought. The equipment scans about a one-eighth-inch-diameter area of the object's surface. The results were fascinating. One of the reproductions from the lot that Henry and I chewed was not "pure" lead; instead, it was found to consist mostly of lead (75.0 percent) plus antimony (14.7 percent) and tin (8.8 percent). The presence of antimony and tin hardens the lead and makes it less compressible. The test musket balls were, in fact, cast from modern lead alloy.

Three musket balls from Monmouth Battlefield were analyzed: a 0.69-inch-diameter dropped British ball (242-9-983), a 0.59-inch-diameter dropped American musket ball (242-9-976), and a 0.61-inch-diameter impacted American musket ball (224-10-281). Table 7.1 shows the major components for each.

None of the patina was removed to expose the musket balls' lead alloy. The areas where the Monmouth Battlefield musket balls were excavated are rich in mineralized iron. The iron is in the patina and was picked up from the soil surrounding the buried musket balls. They are essentially pure lead with no antimony.

Table 7.1. Major elemental analysis of three musket balls from Monmouth Battlefield determined by X-ray fluorescence (XRF)

	Musket balls		
Element	242-9-983	242-9-976	224-10-281
Lead (%)	92.3	92.1	95.6
Iron (%)	6.7	7.0	3.1
Tin (%)	0.0	0.0	0.0

CHEWED MUSKET BALLS

7.13. A reproduction lead musket ball chewed by Dana Linck using both canines and molars. *(Photograph by Dana Linck.)*

When Dana Linck investigated potential chewed musket balls excavated at Fort Montgomery, New York, he also chewed on a reproduction lead musket ball (figure 7.13). He found bite patterns similar to those Henry and I had found in our earlier experiment. Unfortunately, the composition of Dana's reproduction musket ball is not known.

Although these results are interesting, they may not reflect a level of musket ball deformation obtained with pure lead specimens. Therefore, I purchased several bars of 99.9 percent pure lead. Jim Stinson cast a small lot of musket balls for me under controlled conditions. Jim cleaned his melting equipment and mold to ensure that there would be minimal contamination. He cast a dozen 0.67-inch-diameter musket balls (shown in figure 7.14). These cast musket balls were considerably shinier than the previous samples. To distinguish different dentition marks, I chewed on one with only my rear molars, bearing down multiple times with as much force as I could withstand. I used another musket ball to bite with primarily my canine teeth, again as hard as I could endure. Attempts to use my front incisors left very small indentations and are not shown.

7.14. Twelve reproduction musket balls and the mold Jim Stinson used to cast them. *(Photograph by Jim Stinson.)*

I was not very surprised to determine that I was able to make deeper indentations in the 99.9 percent lead musket balls but could not crush them flat. The impressions made by my teeth were much shallower than those shown previously as being swine-chewed.

Figure 7.15 shows the molar-chewed, pure lead reproduction musket ball (right photograph) compared to a musket ball recovered at the Parsonage Farm site at Monmouth Battlefield (left photograph). The latter appears to have been dropped and was found in a location where the British infantry or grenadiers were formed while attacking American general Anthony Wayne's Pennsylvania troops, who had taken a defensive position in an orchard. The diameter of this musket ball is 0.63 inches, which is smaller than a Brown Bess and at first impression would seem to be American. A significant number of impacted 0.63-inch musket balls were found where the orchard stood in 1778, suggesting that they were fired from British fusils. Although the Parsonage Farm was used as a field hospital during the battle, musket ball 224-10-741 was not in the hospital area. The diameter and location where it was found suggest that it was chewed by

MUSKET BALL
AND SMALL SHOT
IDENTIFICATION

7.15. Left: Eighteenth-century musket ball excavated at Monmouth Battlefield State Park at the Parsonage Farm site. *Right:* Reproduction musket ball chewed by the author using only back molars.

a British soldier. Possibly the soldier was wounded during the fight and used a musket ball to bear the pain, but it is more likely that it was chewed to help promote salivation. The level of deformation of this artifact is not that severe. It was unusually hot on Sunday, June 28, 1778. Joseph Plumb Martin described the heat of the day in his memoirs: "It was ten or eleven o'clock before we got through these woods and came into the open fields. The first cleared land we came to was an Indian cornfield, surrounded on the east, west and north sides by thick tall trees. The sun shining full upon the field, the soil of which was sandy, the mouth of a heated oven seemed to me to be but a trifle hotter than this ploughed field; it was almost impossible to breathe. We had to fall back again as soon as we could into the woods" (Martin 1988: 126–27).

7.16. Seventeenth-century round musket ball (*left*) and a human-chewed musket ball (*right*) excavated at the Pope's Fort site, Historic St. Mary's City, Maryland. *(Photograph provided by Henry Miller, PhD; courtesy of Historic St. Mary's City.)*

The depth and shape of indentations on the molar-chewed reproduction are nearly identical to those of the eighteenth-century musket ball shown in figure 7.15. I was able to make relatively deep depressions with the cusps of my molars and could gouge the lead if I pinched it on the side of the ball between the upper and lower molar cusps.

Figure 7.16 shows two musket balls excavated at the early seventeenth-century Pope's Fort site in Historic St. Mary's City, Maryland. The left musket ball has not been used or fired and is shown for comparison to the right musket ball. The denture impressions were identified by Henry Miller as being human. Being found at a military fort, this musket ball may have actually been associated with pain.

I took another 99.9 percent pure lead reproduction musket ball and chewed it with only my left canine teeth and bit as hard as I could. I bit directly into the musket ball to make indentations and bit the side of the musket ball to create gouges. The results are shown in figure 7.17, in which different views of my

7.17. Upper and lower left photographs: Eighteenth-century musket balls excavated at Monmouth Battlefield. *Upper and lower right photographs:* Two views of the same 99.9 percent pure lead reproduction musket ball chewed by the author using only canine teeth.

chewed musket ball are compared to two musket balls excavated at Monmouth Battlefield State Park (the photographs of the reproduction musket ball shown in figures 7.17 and 7.18 are different views of the same musket ball). Musket ball 211-11-005 (upper left photograph) was excavated where the Continental hand-picked troops under Colonel Butler were returning from town and encountered the British Second Battalion of the Forty-Second Regiment of Foot. The area is very similar to the hot, open field described earlier by Martin, and the weather would have been very hot. Musket ball 234-9-836 (lower left photograph) was recovered from the site where a handpicked Continental platoon that included Joseph Plumb Martin chased and fired upon the retreating Forty-Second Regiment. Both artifacts very closely match the dentition impressions in the pure lead reproduction musket ball shown in the upper and lower right photographs of figure 7.17, made using my canine teeth. The two original musket balls were probably used to promote salivation. This would especially be true for artifact 234-9-836. The Americans had fired several rounds at the British. The procedure

MUSKET BALL AND SMALL SHOT IDENTIFICATION

of loading a flintlock musket begins with tearing a paper cartridge open with your teeth. Having personally done this with blank cartridges at many Revolutionary War reenactments, I can attest that black powder often enters your mouth. Black powder is made from saltpeter, charcoal, and sulfur. It tastes salty and dries your mouth and lips. Added to the extreme heat of the day, this could have been stressful for the soldiers at Monmouth. Absent an understanding of the concept of lead poisoning, the idea of chewing on a cool lead ball that would help generate saliva must have seemed a logical decision at the time.

Musket ball WP6-853 (figure 7.18, left photograph) was excavated at Redoubt 4 at Fort Putnam on the Hudson River. Today Fort Putnam is part of the U.S. Military Academy at West Point, New York. The fort was built in 1778 and was never attacked by the British, despite the traitorous attempt by General Benedict Arnold to give the fort's layout and defenses to the British command. Small indentations in artifact WP6-853 match indentations impressed into the 99.9 percent pure lead musket ball by my left upper and lower canine teeth (figure 7.18, right photograph). Artifact WP6-853 seems lightly chewed and may have been used to help with salivation on a hot summer day.

There appears to be much lore about bullets being used to bite on during field surgery conducted without an anesthetic. I have personally chewed on pure lead musket balls to determine the amount of deformation that can occur, but not wanting to become a victim of lead poisoning, I did not spend much time trying to mash them. I also lacked the incentive of severe pain being inflicted upon me while conducting the experiment. I found it very difficult to create much distortion to the lead ball. However, based on the amount of lead having to be displaced, possibly a smaller-diameter musket ball could produce a different effect. There are historical accounts of musket balls being heavily deformed by men under the lash (refer to appendix B for more detail).

 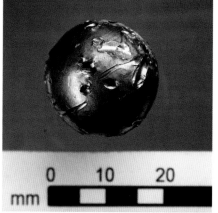

7.18. Left: Musket ball excavated at the U.S. Military Academy at West Point, New York. *Right:* Reproduction musket ball chewed by the author using only canine teeth.

The analysis of human-chewed musket balls is speculative and very much dependent on where the artifacts are found and the known circumstances of why they may have been chewed. I caution the reader that it is often difficult to distinguish the differences between young swine dentition impressions and human molar marks. Just looking at the artifact could lead to the wrong conclusion. The context from which the artifact came from must also be taken into account.

To date, 1,301 musket balls have been excavated at the site of the June 28, 1778, Battle of Monmouth. Of these, 83 (6.4 percent) have been identified as being chewed, with 3.2 percent chewed by a pig or boar; 1.5 percent, human; 1.3 percent, undetermined; and 0.4 percent, rodent (large or small).

CHAPTER 8

"Pewter" Musket Balls

THE VAST MAJORITY OF EXCAVATED MUSKET BALLS that I have observed have a smooth patina, or coating, and the surface features of the shot such as the sprue cut and mold seam are distinctly visible. Occasionally I have seen and excavated musket balls at American Revolutionary War sites that have a blistered surface with a grayish color and sometimes flaking patina. To the untrained eye, the surface characteristics of these alloy musket balls can easily be misidentified as having been chewed. The blistering suggests that this is an alloy that exhibits a light galvanic action when in moist acidic soils. I have observed this type of aging in low-grade (that is, low tin content) pewter ware and buttons. These musket balls, which generally have a slightly lower specific gravity than pure lead, are rarely flattened from impact, suggesting that they are harder than lead.

Lead was in short supply in the colonies. Most of the lead came from Europe, but the French also had some lead mines in America along the Mississippi River and in the Great Lakes region (Burns 2005: 112). One of the few sources of supply in the rebelling colonies was in the western corner of Virginia at Fort Chiswell. Discovered in 1756 by Colonel John Chiswell, these mines supplied lead for the American army during the Revolutionary War, but there was a need to stretch out the lead supply. One way to accomplish this was by mixing lead with other materials of similar or lower melting points. Lead has a melting point of 621.5°F (327.5°C) and can be melted over a campfire. The only other common metal with a melting point low enough to be mixed with lead is tin. Tin has an even lower melting point, 449.5°F (231.9°C), and it alloys well with lead. The only problem was that there were no significant tin mines in North America in the late eighteenth century.

Pewter was much more readily available in the form of plates, cups, tankards, buttons, and so forth. It could be donated to the cause or confiscated from Loyalists. In Monmouth County, New Jersey, for example, the residents were nearly equally split between Patriots and Loyalists. Through the course of the Revolutionary War, a civil war also occurred in the county between these groups, resulting in a large number of houses on both sides being pillaged and burned

(Adelberg 2010). Some of the Loyalist pewter ware could have ended up being melted down for the American cause.

High-quality pewter is approximately 80 percent tin and 20 percent lead. Tin has a density of 7.31 grams per cubic centimeter, whereas the density of lead is 11.34 grams per cubic centimeter. A pure pewter musket ball would have a much lighter weight at a given diameter than the same size lead ball. Therefore, the flight characteristics and force of impact would be different. Pewter could be blended with lead at a low level so as to not greatly change the musket ball's characteristics and would reduce the amount of lead required. This chapter examines the use of pewter being utilized to alloy with lead to make projectiles for eighteenth-century flintlock muskets.

From the four sites discussed in chapter 2, spherical musket balls that were gray and/or blistered, thus visually appearing to be made of lead alloy, were weighed and their diameters measured. These data were used to calculate the metal densities. The results are shown in table 8.1.

Table 8.1. Calculated density of American Revolutionary War musket ball lead/tin alloy

Site	Occupying forces	Density (g/cm^3)	Sample size	Standard deviation	Minimum (g/cm^3)	Maximum (g/cm^3)
Valley Forge—Washington Chapel	American	10.30653	7	0.15981	10.12220	10.51757
Monmouth advance American camp	American	10.23837	5	0.15670	9.96528	10.36605
Battle of Monmouth	British and American	10.27288	78	0.40833	8.00669	10.63921
Neuberger—retreat from Monmouth	British		0			
Overall average		**10.27358**	**90**	**0.38371**		

An overall average calculated density of 10.274 grams per cubic centimeter was obtained for the four sites. It is interesting to note that musket balls recovered at the site occupied only by British troops had no evidence of the lead being alloyed. Based on this density value, a slightly different diameter formula was derived:

Diameter in inches (lead alloy musket ball) = $0.2247 \times$ (weight in grams)$^{1/3}$

By knowing the density of a "pure" lead musket ball and the density of tin, the following formula can be used to calculate the theoretical tin content in the alloy:

$$\text{Alloy density} = \frac{\text{density of lead} \times (100 - X)}{100} + \frac{\text{density of tin} \times X}{100}$$

Where X is the percentage of tin.

The "pewter" musket balls from table 8.1 provide an example of how to calculate the theoretical tin content:

$$10.274 = \frac{10.532 \times (100 - X)}{100} + \frac{7.3 \times X}{100}$$

Therefore $X = 7.98$, or nearly 8 percent tin.

Note that I used the calculated density for pure lead rather than the actual density since the musket balls still contain potential air pockets and impurities.

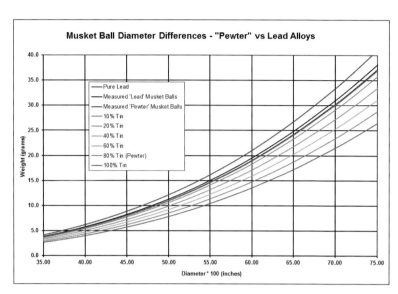

8.1. Musket ball diameter versus weight curves of various lead/tin alloys.

Because true pewter is 80 percent tin, these results show that the so-called pewter musket balls are, in fact, a lead alloy with a small amount of tin. As stated earlier, it appears that pewter objects may have been melted down and blended with lead. An average of 8 percent tin content would indicate that an average of 10 percent pewter was being added to 90 percent lead for the musket balls examined in this study. The higher the tin content, the lighter a given diameter ball would be, as shown in figure 8.1.

A typical American 0.63-inch-diameter musket ball has a weight of 22.6 grams. If it was made of high-grade pewter (80 percent tin), it would weigh only 17.1 grams. The difference of 5.5 grams would undoubtedly produce much different flight characteristics for a given powder charge. Because the force of impact is a function of mass and object acceleration ($F = ma$), the effect on a target would be diminished. Therefore, although it is possible, it is unlikely that high-quality pewter ware objects were melted down for use as shot, but it is possible that pewter could have been blended with lead to stretch short lead supplies.

One of the properties of lead/pewter alloys is that they degrade in the ground over time. The musket balls shown in figure 8.2 were excavated at Monmouth

8.2. "Pewter" alloy musket balls excavated at Monmouth Battlefield State Park.

8.3. "Pewter" alloy American regimental buttons excavated at the Washington Memorial Chapel site, Valley Forge, Pennsylvania. *(Photograph courtesy of the Maryland Archaeological Conservation Laboratory.)*

Battlefield from active farm fields. What caused the surface blistering and scaling? I originally thought that it was caused by the heavily nitrate-based, modern synthetic fertilizers used in these fields. The nitrates can become nitric acid when in solution, causing certain metals to dissolve. However, dissolved metals typically pit, not blister. The lightly acidic soils are also a perfect media for galvanic action, which can cause a blistering effect. Brass artifacts, such as coins, have been excavated that have surfaces so damaged that their dates and origins are nearly impossible to identify, whereas coins found a few yards away in farm lanes that had not been cultivated or fertilized have very readable features.

However, "pewter" or lead alloy buttons found at the Washington Memorial Chapel site at Valley Forge in an undisturbed context were also degraded badly, as shown in figure 8.3. The buttons shown in the photograph were nearly unreadable when excavated but were professionally conserved at the Maryland Archaeological Conservation Laboratory. Although the site was not plowed, it is wooded. The degradation is most likely due to a galvanic effect caused by the lead and tin mixture in soils that are acidic because of tannic acid from the trees and/or acid rain.

Several "pewter" musket balls were tested using the XRF equipment discussed in the previous chapter (table 8.2). The items were tested without removing any surface oxidation so as to not damage the artifact. As stated earlier,

Table 8.2. Major elemental analysis of three "pewter" musket balls determined by X-ray fluorescence (XRF)

	Artifact		
Element	207-12-021	VF10-201	VF10-137 (dice)
Lead (%)	92.9	90.8	96.4
Iron (%)	0.4	5.2	0.3
Tin (%)	5.7	1.8	1.7

MUSKET BALL
AND SMALL SHOT
IDENTIFICATION

8.4. Two facets of a "pewter" alloy gaming die excavated at the Washington Memorial Chapel site, Valley Forge. *(Photograph courtesy of the Maryland Archaeological Conservation Laboratory.)*

8.5. WLG9-579 (*left*) is a lead musket ball, and WLG9-587 (*right*) is a "pewter" alloy musket ball. They were excavated relatively close to each other at the site of the Battle of Cooch's Bridge, Delaware.

iron is from the soils and is in the patina layer. All had a measurable level of tin present.

Artifact 207-12-021, recovered from the Sutfin orchard at Monmouth Battlefield, was identified as a piece of canister shot. Two artifacts, VF10-201 and VF10-137, were found at the Washington Memorial Chapel site at Valley Forge. The first, VF10-201, is a dropped 0.65-inch-diameter musket ball. This is the standard size for a French musket. The second, VF10-137, is a lead alloy gaming dice that appears to have been made from lead and tin (probably from pewter). It was probably made from a musket ball measuring 0.61–0.65 inches in diameter. It was chewed by a pig, so some metal may be missing. Two facets are shown in figure 8.4.

This artifact had very little patination, which would explain the low level of iron found in the XRF analysis. It does, however, show some light surface blistering. The gray color and rough surface together serve as a good indication that the musket ball is not pure lead and probably contains pewter/tin. Figure 8.5 shows two musket balls of nearly the same diameter that were found 232 feet from each other at the site of the 1777 Battle of Cooch's Bridge in Delaware. Ball WLG9-579, on the left, appears to be a standard, lead musket ball with a white/tan smooth patina. Ball WLG9-587, on the right, has the classic gray color, a pitted surface, and little or no patina, indicating that it is a lead/pewter alloy. This suggests that it is most likely an American-made musket ball.

Although the tin content hardens the lead slightly, fired pewter alloy musket balls will deform upon impacting a hard surface. Figure 8.6 shows two impacted musket balls excavated at Fort Montgomery, New York. These musket balls were found in the general area where Americans defending the fort on October 6, 1777, were overrun by the British. Both balls appear to be lead/pewter alloys based on the surface characteristics, and both are deformed from impact. They are most likely American.

8.6. Impacted "pewter" alloy musket balls excavated at the site of Fort Montgomery, New York. *(Photographs by Dana Linck.)*

One of the possible sources for the lead/tin alloy musket balls may have been identified in the historical records of the American Revolution. After the reading of the Declaration of Independence on July 9, 1776, in New York City, the inspired mob marched to the Bowling Green (Battery Park today) and pulled down a gilded, leadened statue of King George III and reportedly used the fragments to produce 42,088 musket balls (CTSSAR 1998; Wolcott 1776). Ebenezer Hazard wrote the following from New York City: "The King of England's Arms have been burned in Philada. & his Statue here has been pulled down to make Musket Ball of, so that his Troops will probably have melted Majesty fired at them" (Gates 1776).

Figure 8.7 is an etching depicting the event; however, the details of the statue in this illustration are inaccurate. The original statue was that of King George on a horse. A fragment of the horse's tail is currently on display at the New York Historical Society in New York City. Several nineteenth-century prints portray George on horseback.

8.7. La destruction de la statue royale a Nouvelle Yorck (The Destruction of the Royal Statue in New York), hand-colored etching, ca. 1776. *(Photograph courtesy of the Library of Congress Prints and Photographs Division, Washington, D.C., repr. no. LC-USZC4-1476.)*

MUSKET BALL
AND SMALL SHOT
IDENTIFICATION

Oliver Wolcott of Litchfield, Connecticut, described this statue in more detail: "An Equestrian Statue of George the Third of Great Britain, was erected in the City of New York on the Bowling Green, at the lower End of Broad Way. most of the materials were lead. but richly Guilded [sic] to resemble Gold. At the beginning of the Revolution, this Statue was overthrown; Lead being then Scarce & dear, the Statue was broken in pieces & the metal transported to Litchfield as a place of Safety: The Ladies of this Village converted the Lead into Cartridges for the Army" (Wolcott 1776). Knowing that fragments of this statue were at the New York Historical Society in New York City, I became interested in the possibility of trying to ascertain whether any of the "pewter" musket balls found at Monmouth Battlefield had come from the statue.

Because lead is very soft for a sculpture, I wondered whether the "leadened" statue was in fact a lead/tin alloy, such that some of the "pewter" musket balls could actually have been cast from statue parts. I speculated that a pure lead statue would be too soft to be self-supporting: it was most likely a lead/tin alloy with at least 15 percent tin, and more likely 20 percent or more, based on the Brinnel Hardness Number (see figure 6.8). The next step would be finding a way to prove this hypothesis.

I attended the 2014 Fields of Conflict Conference in Columbia, South Carolina, and met Michael Seibert, archaeologist at the National Park Service Southeast Archeological Center in Tallahassee, Florida. He gave a paper on X-ray fluorescence (XRF), a nondestructive method used to analyze elemental compositions of various materials. I spoke to Michael about my interest in comparing musket ball lead from Monmouth Battlefield to the King George III statue remnants, and he was very interested. He secured funding through the National Park Service Outreach program to do the testing on site. Thanks to the help of Scott Wixon, collections manager at the New York Historical Society, I learned that there were actually six pieces of the statue at the New York Historical Society Museum: two were on display and four were in storage. Scott arranged for us to have access to the latter for XRF testing. On September 24, 2014, Michael and I met Scott at the museum, and he brought us the two small and two very large pieces of the statue for analysis (figure 8.8). Scott stated that the fragments were acquired in 1875, 1878, and 2001 and that all were from Connecticut.

The large piece in the foreground of figure 8.8 has cut marks that appear to be from an axe and a large square hole that is most likely from a pick (figure 8.9). These markings suggest that the statue was hacked to pieces by an angry mob, as described by Wolcott.

Michael proceeded with the analysis with a Bruker Tracer series pXRF spectrometer, as shown in figure 8.10. He tested all four statue fragments with multiple locations on the large pieces. He began by testing the outer surface area of the statue.

8.8. Fragments of the gilded, leadened statue of King George III.

8.9. Statue fragment having cut marks (*left*) and a pick hole (*right*).

8.10. Michael Seibert testing large statue fragment using X-ray fluorescence (XRF) spectroscopy.

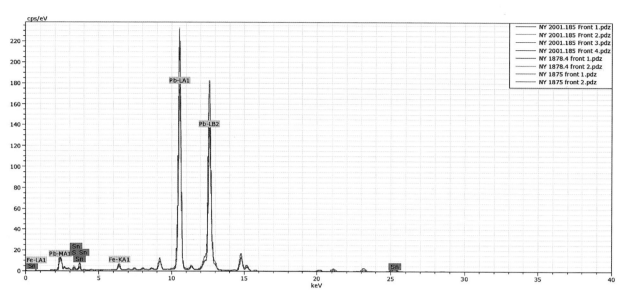

8.11. Spectrum of the exterior surface of the statue fragments showing the low tin (Sn) levels.

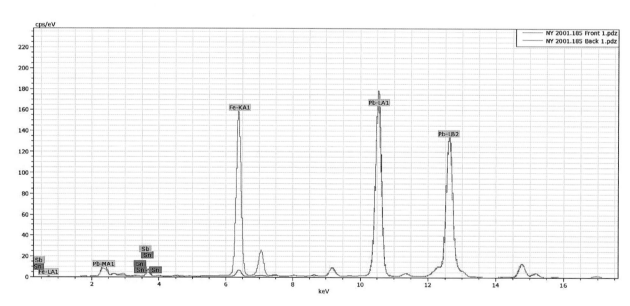

8.12. Spectrum of the exterior (front) and interior (rear) surfaces of one of the statue fragments showing the different iron (Fe) values obtained.

8.13. Iron wire frame in the reverse side of a statue fragment.

The preliminary results were surprising. We expected 20 percent tin (Sn) or more, and the test results were typically less than 5 percent, as shown in figure 8.11.

I did not think a hollow lead statue would support itself under the tremendous weight. However, when Michael began testing the reverse side, or the inside, of the statue, we had another surprise: very high iron (Fe) content readings, as shown in figure 8.12.

Close examination of the inside layer of the statue revealed the outline of a support frame under the lead (Pb) coating. The statue was lead over an iron mesh (figure 8.13) and bar frame (figure 8.14). Michael, I, and the museum staff were amazed. The existence of this support frame was not known before the analysis. The museum staff were happy to learn that although in the past this very small

8.14. Iron support bar inside statue fragment. This fragment was most likely from an appendage of either the horse or King George.

MUSKET BALL AND SMALL SHOT IDENTIFICATION

fragment was too small to positively identify as truly being from the statue based on its shape and size, the XRF analysis of the elemental metallic composition for this piece was identical to all of the other statue fragments.

Figure 8.14 also shows that the edges of the fragments are curled from being melted. This helps to support the story of the statue being melted down to make musket balls. These fragments must have been kept in hiding. One must put this into the proper perspective: the owners of these fragments put themselves at risk just to have them. If caught by the British army, they would most likely have been executed for committing a treasonous act directly against the king!

On September 25, 2014, Michael and I went to Monmouth Battlefield State Park to analyze 104 musket balls that had blistered patinaed surfaces. Michael tested several pieces by measuring through the patina and then scraping the same section to shiny metal and testing again. He observed that the tin content did not change. As seen in earlier tests, only the percentage of iron changed, and this was attributed to the local soils having high iron contents, which accumulated in the lead carbonate patina. Of the 104 musket balls, 10 had tin levels consistent with the tin levels of the statue fragments. Figure 8.15 is a close-up view of the tin peaks of a representative spectrum of one of the matching musket balls.

Ten musket balls excavated at Monmouth Battlefield State Park, all having smooth patinaed surfaces, and 3 freshly cast reproduction musket balls manufactured from 99.9 percent pure lead were sent to Michael for testing as to establish baselines. None of the 10 smooth patina musket balls or the 99.9 percent pure lead reproductions had any significant levels of tin. It is interesting to note that 8 of the 10 control musket balls had diameters of 0.69–0.70 inches, suggesting that they were British. However, data for the 104 blistered samples from the

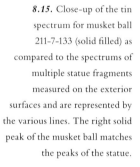

8.15. Close-up of the tin spectrum for musket ball 211-7-133 (solid filled) as compared to the spectrums of multiple statue fragments measured on the exterior surfaces and are represented by the various lines. The right solid peak of the musket ball matches the peaks of the statue.

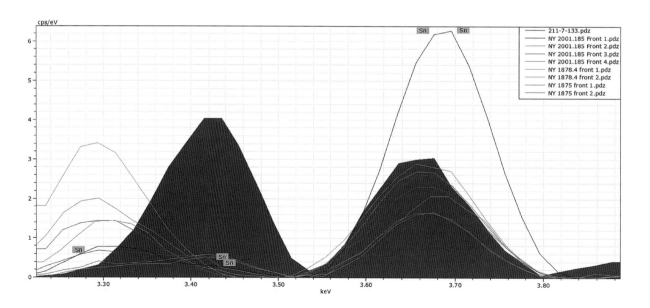

Battle of Monmouth indicate that at least 10 of these lead shot had XRF signatures closely resembling those of the King George III statue fragments. Thus, the British apparently *did* have "melted Majesty" fired at them. The remaining 94 musket balls with blistered patina tested positive for tin but did not match the signature of the statue lead. Therefore, it can be concluded that musket balls with blistered patina are alloys probably containing tin. It also appears that the practice of using tin alloys was restricted to the Americans.

The XRF analysis produced yet another surprise. Three musket balls from Monmouth Battlefield had approximately 30 percent or more tin content. This high level of tin suggests that low-grade household pewter ware may have actually been melted down to make musket balls. These three musket balls had diameters of 0.56 inches, 0.58 inches, and 0.59 inches, indicating that they were most likely from rifles. This is interesting, since the Continental army was typically supplied with premade cartridges. Cartridge making was a cottage industry. Oliver Wolcott lists how many cartridges each of "The Ladies of this Village" made from the statue lead, for a total of 42,088 (Wolcott 1776). Rifles, in contrast, were typically individually made for a specific customer and came with a musket ball mold to match the bore. Therefore, riflemen had to produce their own shot, because of the uniqueness of their weapons. The presence of three musket balls with high tin content suggests that these men were scavenging for lead and were substituting pewter or blending lead with pewter.

Overall the results also suggest that XRF analysis might be able to establish a fingerprint for specific lead shot. It may be possible to use this data to track individual riflemen's movements across a battlefield.

Pewter was being melted and used by the new United States of America to cast items such as buttons. The iconic symbol for this new nation is shown in figure 8.16 on the very rare pewter button found at the Washington Memorial Chapel site at Valley Forge, Pennsylvania, by BRAVO. Note the "1777" under the "USA" letters.

"PEWTER" MUSKET BALLS

8.16. Pewter button with raised "USA / 1777" excavated at the Washington Memorial Chapel site, Valley Forge.

CHAPTER 9

MUSKET BALLS ALTERED FOR NONLETHAL USE

LEAD IS A MALLEABLE METAL and can be easily hammered, carved, inscribed, melted, molded, or otherwise altered. Many soldiers modified musket balls in creative ways for nonlethal uses, producing a variety of items ranging from gaming pieces to pencil lead to toys and sinkers. Note that the diameters shown in photographs in this chapter are the calculated diameters of the original spherical musket balls before alterations; these diameters assume no lead loss.

GAMING PIECES—DICE

Dice are among the oldest gaming implements known to man, and soldiers have been gaming for centuries. According to the King James Bible's account of the crucifixion of Christ, Roman soldiers "crucified him, and parted his garments, casting lots" (Matthew 27:35). Lots are made of bones, wood, or small rocks, and the idea of casting lots is similar to that of throwing dice. Gaming can lead to controversy, which can lead to disagreements and fights. General George Washington, commander in chief of the American forces during the Revolutionary War, was opposed to gaming by the soldiers. On May 8, 1777, he issued the following general order (fig. 9.1):

> As few vices are attended with more pernicious consequences, in civil life; so there are none more fatal in a military one, than that of GAMING; which often brings disgrace and ruin upon officers, and injury and punishment upon the Soldiery: and reports prevailing, which, it is to be feared are too well founded, that this destructive vice has spread its baneful influence in the army, and, in a peculiar manner, to the prejudice of the recruiting Service,—The Commander in chief, in the most pointed and explicit terms, forbids All officers and soldiers, playing at cards, dice or at any games, except those of EXERCISE, for diversion; it being impossible, if the practice be allowed, at all, to discriminate between innocent play, for amusement, and criminal gaming, for pecuniary and sordid purposes. (Washington 1777–78)

9.1. May 8, 1777, general order from Washington, from the George Washington Papers at the Library of Congress. *(Courtesy of the Library of Congress, Manuscript Division.)*

On May 26, 1777, Washington issued General Order No. 349 to Brigadier General Smallwood (fig. 9.2). The following is an excerpt from that document:

> Let Vice, and Immorality of every kind, be discouraged, as much as possible, in your Brigade; and as a Chaplain is allowed to each Regiment, see that the Men regularly attend divine Worship. Gaming of every kind is expressly forbid, as the foundation of evil, and the cause of many [a] Gallant and Brave Officer's Ruin. Games of exercise, for amusement, may not only be permitted but encouraged.
>
> These instructions you will consider as Obligatory, unless they should Interfere with General Orders. Which you must always endeavor to have executed in your Brigade with Punctuality. (Washington 1777–78)

On January 8, 1778, Washington wrote from his headquarters in Valley Forge, "The Commander in Chief is informed that gaming is again creeping into the Army; in a more especial manner among the lower staff in the environs of the camp. He therefore in the most solemn terms declares, that this Vice in either Officer or soldier, shall not when detected, escape exemplary punishment; and to avoid discrimination between play and gaming forbids Cards and Dice under any pretence whatsoever" (Washington 1777–78). In this excerpt, General Washington specifically complains about "Cards and Dice" being used

MUSKET BALL
AND SMALL SHOT
IDENTIFICATION

9.2. May 26, 1777, general order from Washington to Israel Putnam and William Smallwood, from the George Washington Papers at the Library of Congress. *(Courtesy of the Library of Congress, Manuscript Division.)*

by soldiers at Valley Forge. Three complete dice excavated at the Washington Memorial Chapel site in Valley Forge are shown in figure 9.3 on page 131. The photographs show each facet of the dice. Based on their weights, the dice appear to have been made from hammered musket balls. Talk about loaded dice! Artifacts VF7-726 and VF10-137 have weights equivalent to that of 0.61-inch-diameter musket balls. The weight of artifact VF8-1036 matches that of a 0.62-inch diameter ball.

What is interesting about VF7-726 and VF10-137 is that they feature Roman numerals rather than the customary dots. However, the number four on VF7-726 is Arabic. These dice could possibly have been made by carpenters or men who worked on building construction. What suggests this? Houses and barns in the eighteenth century were made from timbers using mortise and tenon joints. Each joint was hand-cut, making it unique and having only one ideal fitting member. To ensure that mated pieces were properly assembled, each mortise and tenon was numbered. This was typically done by chiseling the same number onto each matching pair. The curved nature of Arabic numbers (apart from one and four) makes them difficult to chisel. Roman numerals consist of straight lines and so were preferred by most carpenters. I have observed Roman numeral sets carved in joints at several historic buildings.

MUSKET BALLS
ALTERED FOR
NONLETHAL USE

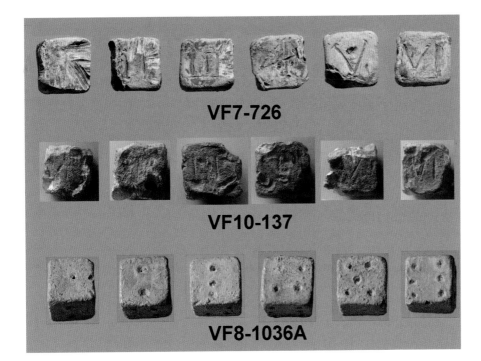

9.3. Three different gaming dice made from musket balls excavated at the Washington Memorial Chapel site, Valley Forge, Pennsylvania. VF7-726 (*top row*) and VF10-137 (*middle row*) have Roman numerals rather than the usual dots. VF10-137 is made from a "pewter" alloy.

It is also interesting to observe that VF7-726 was chewed by a rodent, probably a rat, and VF10-137 was chewed by a swine. The composition of VF10-137 was mentioned in the previous chapter as having 1.7 percent tin, thus making it a "pewter" alloy. The difference in color and low level of patination can be used as a visual identification.

How do we know that dice were being made from musket balls? Musket balls were readily available and could easily be hammered into cubes. The calculated diameters of the theoretical musket balls are consistent with the ordnance found at the site and with the type of French musket being used. Figure 9.4 illustrates the steps required to manufacture a gaming dice from a musket ball by compressing it between two flat surfaces at a time, which produces opposite flat sides on the once-round musket ball. This can be accomplished with a hammer or even a flat rock.

Several dice-in-the-making (two of which are shown in figure 9.5) were excavated at the Washington Memorial Chapel site at Valley Forge. Artifact VF9-029, like the reproduction musket ball number 3 in figure 9.2, has only four flat surfaces. Artifact VF8-973 is a nearly complete cube without any numbers or dots, but the original spherical shape is easily identified in the circular flat surfaces of the artifact. This is identical to reproduction musket ball number 4 in figure 9.2.

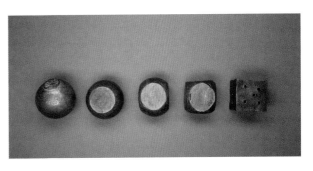

9.4. Reproduction musket balls showing the transformation from a musket ball to a gaming die.

MUSKET BALL
AND SMALL SHOT
IDENTIFICATION

9.5. Two unfinished gaming dice made from musket balls excavated at the Washington Memorial Chapel site, Valley Forge.

Why were these dice and other gaming pieces left behind? Were they simply lost, or were they discarded? One possible explanation is that the Continental army left Valley Forge rapidly to pursue the Crown forces who left Philadelphia and headed to New York. Another, more likely, explanation comes from Washington's January 8, 1778, general order (fig. 9.6), which specifically states, "The Commander in Chief . . . in the most solemn terms declares, that this Vice in either Officer or soldier, shall not when detected, escape exemplary punishment" (Washington 1778–78).

On January 28, 1778, Washington further ordered that each soldier be physically checked for wasting cartridges: "As there has been an extraordinary and unaccountable waste of Ammunition in many of the brigades; as soon as the brigades are completed therewith agreeable to yesterdays [*sic*] orders, the Commander in Chief positively orders . . . the Colonels and Commanding Officers of Regiments to see that an Officer in each Company carefully examine their men's ammunition every day at roll call in the morning and severely punish any soldier who shall carelessly waste a single cartridge" (Washington 1777–78).

This order could possibly have been Washington's method for determining who was using musket balls for gaming pieces, among other uses.

Anyone found to be missing cartridges would be severely punished. The punishment for destroying ammunition was explicit: "That all Non-Commissioned Officers and Soldiers convicted, before a Regimental Court-Martial, of stealing, embezzling, or destroying Ammunition, Provisions, Tools, or any thing belonging to the publick stores[:] if a Non-Commissioned Officer, to be reduced to the ranks, punished with whipping, not less than fifteen nor more than thirty-nine lashes, at the discretion of the Court-Martial; if a private Soldier, with the same corporal punishment" (Washington 1775: 1159).

It is possible that many a gaming piece was tossed out of the huts to avoid such severe punishment, first for gaming and second for destroying ammunition.

9.6. January 28, 1778, general order from Washington, from the George Washington Papers at the Library of Congress. *(Courtesy of the Library of Congress, Manuscript Division.)*

GAMING PIECES—PROBABLE DICE-IN-THE-MAKING

Several musket balls that have been hammered once, making two flat spots like reproduction #2 in figure 9.4, have been excavated at the Washington Memorial Chapel site at Valley Forge. These could have been used as is for games such as checkers, chess, and Parcheesi but could also have been dice-in-the-making, as Washington was specifically concerned about dice and card playing. Figure 9.7 shows a sampling of these artifact types.

The smooth, flat surface on one side of artifact VF9-023 (figure 9.7, bottom row, far left photograph) indicates that the musket ball was struck with a smooth object such as a hammer or a flat rock. The encampment site is near the Schuylkill River, which has many very flat river rocks. The other side of the same artifact (figure 9.7, top row, left photograph) has a rough surface. This indicates that it was smashed either with or against a coarse-surfaced rock, probably used as an anvil. Musket ball VF7-976 was struck unevenly. The coloration and low level of patina of artifact VF7-1102 suggest that it might contain a low level of pewter. It was struck with a very smooth surface.

9.7. Four possible unfinished gaming dice made from musket balls excavated at the Washington Memorial Chapel site, Valley Forge: artifact VF9-023 (*top row, left photograph; top row, right photograph*); artifact VF7-976 (*middle row, left photograph; middle row, right photograph*); artifact VF7-902 (*bottom row, left photograph*); and artifact VF8-961 (*bottom row, right photograph*).

Gaming Tokens

Many board games, such as checkers and Parcheesi, require flat tokens that are used as the gaming pieces. Any games that involved betting required some sort of currency, either actual money or a promissory piece representing an amount in currency. The latter led to the use of the chip, which could eventually be exchanged for the amount of the currency. Currency does not necessarily mean money but could also be possessions of the persons playing. During World War II, for example, cigarettes were the common currency among prisoners of war.

Lead, being very malleable, could be easily fashioned into simple gaming tokens. Many circular and rectangular lead tokens were recovered by Historic Shipwrecks, Inc. from the pirate ship *Whydah* (figure 9.8). The *Whydah* sank during a storm in 1717 off Cape Cod, Massachusetts, and is the only actual pirate ship with "treasure" on board to be excavated to date (Kinkor, Simpson, and Clifford 2007: 8). As can be seen in the drawings, each of these tokens has a telltale "X" (or plus sign) inscribed in the lead.

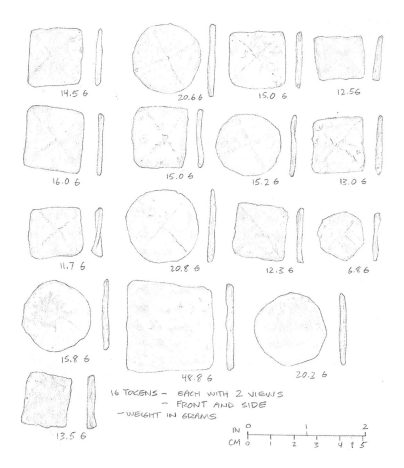

9.8. Sketches of lead gaming tokens recovered from the pirate ship *Whydah*. (Sketches courtesy of Ken Kinkor and the Whydah Pirate Museum, Provincetown, Massachusetts.)

MUSKET BALLS ALTERED FOR NONLETHAL USE

MUSKET BALL
AND SMALL SHOT
IDENTIFICATION

9.9. A possible gaming token (probably made from a flattened musket ball) excavated at the Washington Memorial Chapel site, Valley Forge. The right photograph is a close-up of an "X" inscribed in the center.

A similar token was found at the Washington Memorial Chapel site in Valley Forge (figure 9.9). The right photograph is a close-up of the center and shows the same characteristic "X" cut into the lead. The deep lines cut into the token were made with a blunt metal blade such as a tomahawk or chisel. The meaning of these lines is not known. This token weighs 24.2 grams, which means that if it was made from a musket ball, the ball would have had an original diameter of 0.65 inches. This is consistent with the musket balls found at the site and the size used in a French musket.

Flat lead disks have been found at other sites as well. Figure 9.10 shows two lead disks that were excavated at the site of the 1778 Battle of Monmouth.

The precise uses of these disks can only be speculated about. Their weights suggest that they were made by flattening musket balls. Artifact 224-10-716 (figure 9.10, left photograph) has a weight of 23.2 grams, which would be

9.10. Possible gaming tokens found at Monmouth Battlefield State Park that were probably made from musket balls. Artifact 242-9-977 has a square hole in the center that appears to have been made by a hand wrought nail.

equivalent to a 0.64-inch-diameter musket ball. Artifact 242-9-977 (figure 9.10, right photograph) has a weight of 29.9 grams, which suggests that it was made from a 0.69-inch-diameter musket ball. This artifact also has a square hole punched in the center, which was probably made by a square, hand-wrought nail. This also suggests a possible eighteenth-century origin. Although they are not inscribed with an "X," their overall size and shape are consistent with the *Whydah* and Valley Forge examples. It is interesting to note that the Battle of Monmouth was the first engagement for the Continental army after the Valley Forge encampment. The same soldiers were present in both locations, so the ammunition recovered at both sites are from the same sources.

Several musket balls found at the Washington Memorial Chapel site in Valley Forge appear to have worm marks on them (figures 9.11, 9.12). The marks are in the center of a flat spot as if the musket balls had been intentionally struck with a flat smooth object. Were these used as gaming pieces?

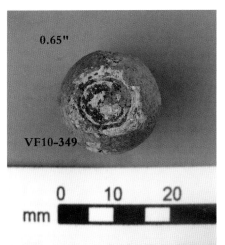
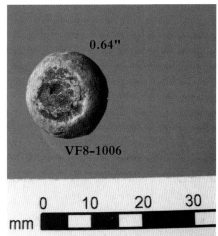

9.11. Two musket balls with worm marks excavated at the Washington Memorial Chapel site, Valley Forge, that may have also been used as gaming tokens.

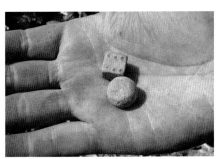

9.12. Two views of a musket ball with worm marks and a flat spot excavated at the Washington Memorial Chapel site, Valley Forge, found next to a gaming dice (*right*).

MUSKET BALL AND SMALL SHOT IDENTIFICATION

9.13. Partially flattened musket ball excavated by Glen Gunther at an American camp site in New Jersey.

9.14. Flattened musket ball excavated at the Washington Memorial Chapel site, Valley Forge.

9.15. Shaped musket ball excavated at the Washington Memorial Chapel site, Valley Forge.

Partially flattened musket balls with circular markings on the flat face have also been found at other sites. Figure 9.13 shows an excellent example found in a Revolutionary War American camp site in New Jersey.

Musket balls that have been flattened more than would have been appropriate for making dice have also been excavated at the Valley Forge site. Figure 9.14 shows a musket ball that was originally 0.69 inches in diameter and was hammered into a flat disc. This was possibly used as a checker or for a similar board game.

Boredom is a common affliction among soldiers, including the ones at Valley Forge. To pass the time, soldiers would become creative in their activities, including turning musket balls into dice and other gaming pieces. Figure 9.15 shows a musket ball that was partially flattened before symmetrical ridges were pressed into the circumference of the ball. The flat spots insinuate that it was made to sit on a flat surface, suggesting that it was possibly a gaming piece such as for chess.

Lead Pencils

Modern "lead" pencils actually use graphite as the media to write with, but elemental lead has been used as a writing implement for centuries. This is why the modern graphite pencil has been misnamed the "lead" pencil (PaperMate 2013). With lead being readily available to the common eighteenth-century soldier, it made a convenient writing tool. William Calver and Reginald Bolton found pencils at British camp sites in New York City, and they describe one rectangular piece of lead among several others as follows: "The other objects . . . are derived from bullets, and reflect the whims of the British soldier, his pastime or his needs. The rounded end of number 8 betrays the bullet origin of this 'lead' pencil" (Calver and Bolton 1950: 77–78).

Seventeen lead pencils have been excavated at Monmouth Battlefield. Figure 9.16 is an example of a lead pencil that, from the folds in the lead, appears to have been made from a musket ball probably larger than 0.57 inches in diameter (assuming lead loss from use).

Just because an artifact appears to be a lead pencil, one cannot assume that it was made from a musket ball. Figure 9.17 is a lead pencil that appears to have been bent by agricultural equipment. It was found in Newark, Delaware, at the site of the September 1777 Battle of Cooch's Bridge. The British took the field and camped there for several days. This artifact was excavated in what appears to be one of the British camp sites, based on associated artifacts found in the same place (buttons, buckle, coins, and so forth). It weighs only 9.0 grams, and there do not appear to be any fold lines. It could have been made from a

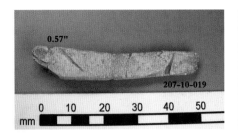
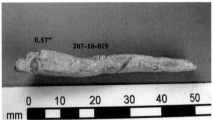

9.16. Two views of a lead pencil found at Monmouth Battlefield State Park, probably made from a 0.57" diameter or larger musket ball.

0.46-inch-diameter rifle ball but was more than likely made from a piece of scrap lead.

Toys

As noted earlier, bored soldiers become very creative in making things to keep themselves occupied or amused. Figure 9.18 shows two lead "whizzers" that may have been made from musket balls. These are operated by passing a string through both holes and tying it off into a loop. The loop ends are placed around a finger on each hand, and the whizzer is twirled to twist the string. Then, when the strings are pulled back and forth like an accordion, the whizzer spins at a high speed and makes a buzzing sound. As a child, I did the same thing using a large coat button. Listening to the noises provided a simple and almost hypnotic effect, and there was great satisfaction in seeing how tight I could get the string to twist. The artifact in the left photograph in figure 9.18 was found at a British camp site in New York City in the early twentieth century by Calver and Bolton. They assumed that it was made from a musket ball. The artifact is currently in the collections of the New York Historical Society. Calver and Bolton reported that it was made for children in the camp (Calver and Bolton 1950: 78). There is no way to substantiate that claim, however, and it may have

9.17. A lead pencil, possibly damaged by agricultural equipment, found at the site of the 1777 Battle of Cooch's Bridge in Newark, Delaware.

9.18. Whizzers: *left*, excavated by the Field Exploration Committee at a British camp site in New York City (*Collection of the New-York Historical Society, INV.5925*); *right*, excavated at Fort Niagara, Youngstown, New York (now in the collections of the Rochester Museum and Science Center, Rochester, New York).

been for a soldier. The similar artifact in the right photograph in figure 9.18 was excavated at Fort Niagara (1759–1812) in Youngstown, New York.

Fishing Sinkers

Lead has been used to produce fishing sinkers and net weights for centuries. It has ideal characteristics for the purpose: a very high density but malleable and easily molded. This is an account of lead being used in ancient Egypt: "As depicted on the walls of Egyptian temples and tombs, large-scale fishing consisted of well-coordinated efforts of several people, who joined forces and used a seine-net (a large net with cork floats along the top edge and weights along the bottom edge). The net was weighted down with lead (an example of these lead weights . . . [is] presently housed at the Berlin Museum)" (Gadalla 2007: 106).

Five lead net sinkers were excavated at a late eighteenth- to early nineteenth-century farmhouse site at Monmouth Battlefield State Park. The large hole is to accommodate coarse fishing-net twine. The hole is large and very uniform in diameter, suggesting that it was made during the casting process. These lead sinkers are shown here (figure 9.19) because their shape and size could lead to their being erroneously identified as having once been musket balls or artillery shot. Fortunately, two brass rowboat oarlocks were also recovered from the same site, which aided in accurate identification.

Musket balls that had holes drilled completely through were recovered from the 1717 wreck of the *Whydah* off of Cape Cod (figure 9.20). These were most likely used as fishing sinkers. The holes are much smaller than those in figure 9.19, and the overall shapes of the shot are more irregular. One of the musket balls in figure 9.20 has a square hole, which may have been made by a small nail. The lead shot with larger holes were possibly used as net sinkers.

Figure 9.21 shows a musket ball that is described in the site report as having "a hole drilled through it." It was excavated by Steve Warfel at the Ephrata Cloister site in Mount Zion, Pennsylvania. The cloister dormitory was occupied by part of the Continental army for seven months during the winter of 1777–78 and was used as a hospital. Warfel reports that melted lead "casting sprue" was excavated, and he therefore speculates that "[r]ecuperating soldiers were . . . assigned the task of making lead balls before their discharge and return to combat unit" (Warfel 2001: 22–23). The hole in the musket ball shown in figure 9.21 was obviously not a casting flaw but shows intentional alteration to the ball. This was most likely intended to be a fishing sinker. Both of the

9.19. Late eighteenth- to early nineteenth-century lead net sinkers. *(Photographs by Michael Timpanaro.)*

9.20. Musket balls converted into lead sinkers by pirates. *(Photograph courtesy of Chris Macort and the Whydah Pirate Museum, Provincetown, Massachusetts.)*

9.21. Two views of a musket ball excavated at the Ephrata Cloister, Mount Zion, Pennsylvania, showing both sides of a hole that completely passes through the ball. *(Photograph courtesy of the State Museum of Pennsylvania, Pennsylvania Historical & Museum Commission.)*

shaft entrances are tapered inwardly. It is difficult to determine how this was produced. It may have been simply carved with a tip of a knife to prevent the edges of a nontapered hole from abraiding a fishing line. It could also be a wear pattern from being on a string. This is very similar to the musket ball in the upper left of figure 9.20.

Figure 9.22 shows an obvious lead sinker that was excavated at a 1777–78 Continental army encampment site at Valley Forge. This site had hut footprints very close to the Schuylkill River. This artifact, in conjunction with other fishing-related artifacts, indicates that the soldiers were supplementing their meager rations with fish. This sinker was made by hammering the lead to shape (as opposed to being cast from molten lead), as can be seen by the multiple flat impact marks. This suggests that it was made from a 0.55-inch-diameter musket ball (assuming no lead loss), since numerous altered musket balls were recovered at this site but no lead bar stock was found. A fine wear pattern of a fishing line is also present at the top of the hole.

9.22. Lead sinker excavated at the Washington Memorial Chapel site, Valley Forge.

Melted Musket Balls

Partially melted musket balls have been found at several Continental army camp sites. Figure 9.23 shows one example of three partially melted musket balls found at the 1777–78 American camp site at the Washington Memorial Chapel site at Valley Forge. The photograph on the left shows that this ball was cast and the casting sprue was cut off before it was remelted. There are no obvious indications that it had any defects that would have caused it to be rejected. The deformation suggests that the musket ball may have been on a hot surface, such as a rock in or

MUSKET BALL
AND SMALL SHOT
IDENTIFICATION

9.23. Two views of a partially melted musket ball excavated at the Washington Memorial Chapel site, Valley Forge.

near a fire. Figure 9.24 is a partially melted musket ball found at a June 1778 Continental army camp site in Englishtown, New Jersey. Troops camped at this site the day before, and possibly the night after, the Battle of Monmouth.

Unidentifiable Lead Objects

Several unidentifiable lead objects have been recovered at the chapel site at Valley Forge. Figure 9.25 shows two different artifacts with nearly identical shapes. They look like wheels for a child's toy, but there are no axle holes. Artifact VF9-008, in the right photograph, has had the lead flashing sculpted out with a knife, and the top spoke is cut through. The residual flashing indicates that these artifacts were made in a crude mold.

9.24. Partially melted musket ball excavated at a Continental army camp site in Englishtown, New Jersey.

9.25. Two lead "wheels" excavated at the Washington Memorial Chapel site, Valley Forge.

One of the more intriguing artifacts found at the Valley Forge Continental army winter encampment site is shown in figure 9.26. It appears to be a sad or disgruntled human face carved into a partially flattened musket ball. There appears to be a tear coming out of one of the eyes. Is this a depiction of an American soldier enduring a harsh winter at Valley Forge, or is it the face of the enemy for whom this ball was intended?

9.26. Musket ball with a carved "face" recovered from the Washington Memorial Chapel site, Valley Forge.

Modern Bullets

Modern bullets are often found in earlier military areas. Because of the higher velocity of newer weapons, their bullets drastically deform and can appear to be musket balls from an earlier period. They could have been fired by a hunter or sportsman years or decades after the military engagement. The objective of showing them in this book is to help the reader in identifying them properly. Figure 9.27 shows an artifact found in a high-conflict area of Monmouth Battlefield

9.27. Two views of a "modern" bullet found at Monmouth Battlefield that looks like an impacted musket ball.

MUSKET BALLS ALTERED FOR NONLETHAL USE

MUSKET BALL AND SMALL SHOT IDENTIFICATION

9.28. Two views of a modern bullet excavated at the Washington Memorial Chapel site, Valley Forge.

State Park. Initially it looks like a musket ball (left photograph); however, the right photograph showing the other side shows the raised ring that was the base of the modern projectile. The weight (27.7 grams) and diameter indicate that it is most likely a nineteenth-century Minié ball.

Figure 9.28 shows an even more modern, probably twentieth-century, bullet excavated at the Valley Forge encampment. This object is somewhat more easily identified, because part of the back of the conical bullet is still intact. This is approximately a .40 caliber bullet.

CHAPTER 10

SMALL SHOT... NOT JUST FOR THE BIRDS

AS STATED IN EARLIER CHAPTERS, small lead shot could be used by itself in a gun (thus the later term "shotgun") or with a musket ball. Both were designed to improve lethality with a widespread pattern but at a short distance. Figure 2.6 lists a variety of sizes of small shot found at the Jamestown fort site. Historically, small shot was used for hunting or fowling. Small shot was named for the type of game that was being hunted, such as pigeon, snipe, plover, partridge, duck, gray goose, white goose, swan, Bristol, coon, muskrat, beaver, and small and large buck (Potter and Hanson 2001). It falls into three general sizes: buckshot, bird shot, and rat or dust shot (Potter and Hanson 2001). The specific sizes and descriptions of types of shot are given in appendix C. Naming the groups came from the shot's intended use. Buckshot is for large game (primarily deer), as opposed to smaller bird shot, which is for waterfowl and other winged game. Rat shot is used for pest control. Because most battlefield archaeological surveys are conducted using metal detectors, very few examples of bird or rat shot are found, because their small size makes them not easily detectable. However, these types of shot have been recovered using classical archaeological procedures at military sites. Figure 10.1 shows examples of bird shot, buckshot, and various sizes of musket balls excavated at the site of Fort Stanwix in Rome, New York. The fort was built by the British in 1758 but was used by the Continental army throughout the Revolutionary War until most of it burned down in 1781.

10.1. Examples of bird shot (*top row*), buckshot (*middle row*), and musket balls (*bottom row*) on display at Fort Stanwix National Monument, Rome, New York.

We know from earlier chapters that the military used both ball and buckshot during the Revolutionary War. Buckshot has changed little in the past two centuries. Shot excavated at Monmouth Battlefield ranged from 0.27 inches to 0.38 inches in diameter. However, it is very difficult to distinguish eighteenth-century buckshot from modern buckshot that has been in the ground for only a few years but has developed a patina. Farms that make up the current park allowed deer hunting until very recently. Because of the potential contamination by modern buckshot, Monmouth Battlefield State Park now allows only rifled shotgun slugs for deer hunting. The primary technique of dating small shot is by identifying the composition of the shot and/or the method of manufacturing. Modern lead shot is machine made and often copper coated. Methods for making small shot evolved over time to increase the manufacturing speed and quantity produced.

Small Shot Made from a Mold

Some early buckshot was made in gang molds. Figure 10.2 shows buckshot still attached together as it came out of a gang mold. These artifacts were excavated at the 1655–75 Seneca village Dann site near Rochester, New York.

10.2. The bar of lead buckshot (*upper left corner*) was made in a gang mold. It was found at the 1655–75 Dann site, a Seneca Indian village in Honeoye Falls, New York, and is now in the collections of the Rochester Museum and Science Center, Rochester, New York.

Mold-cast small shot have the same characteristic sprue cut, and sometimes a mold seam visible, as musket balls, provided they are not too flattened from impact. Figure 10.3 shows two examples of buckshot with these characteristics excavated at Monmouth Battlefield (left photograph) and one excavated at the Washington Memorial Chapel site at Valley Forge (right photograph). Note that the medial ridges are made from clipping off the sprue and coincidentally align with the mold seams in the specimens shown.

One must bear in mind that these small shot were packed as a single load into a musket barrel. Unlike musket balls, imperfections in casting buckshot such as flashing or offset mold halves did not make as great a difference in the barrel's diameter. There was no problem with windage or fouling. Therefore, small shot could be made more crudely than musket balls, which require tighter tolerances.

10.3. Eighteenth-century mold-cast buckshot with visible mold seams and sprue cuts with medial ridges: *left*, excavated at Monmouth Battlefield State Park, New Jersey; *right*, recovered at the Washington Memorial Chapel site, Valley Forge, Pennsylvania.

RUPERT METHOD

Casting lead shot in a mold does have a specific restriction: the diameter of the hole that the lead is poured into must be smaller than the diameter of the shot. Therefore, casting bird shot becomes problematic, because the hole also has to be large enough so that the lead can flow in and air can escape.

I was involved in a project that gave me a very good understanding of the Rupert method of making shot. In July 2010, a thirty-two-foot fragment of a wooden sailing vessel was found in the subsoils of the World Trade Center site in Lower Manhattan, New York. A total of 346 artifacts were found in the ship fragment that had the potential of being eighteenth-century military. I was retained by the archaeological company responsible for the project (AKRF, Inc., Environmental, Planning, and Engineering Consultants) to identify and possibly determine the age of the artifacts. Of these artifacts, 251 were identified as bird shot—lead shot with diameters ranging from 0.08 inches to 0.20 inches inclusively. Only 8 buckshot were found. None of the small shot recovered had any visible mold seams or the telltale clip mark where the sprue had been attached to the ball. They were all "apple" shaped and had a concave dimple, as shown in figure 10.4.

10.4. Bird shot with concave dimples recovered from a wrecked sailing vessel at the World Trade Center site in New York.

In an attempt to identify the type and rough time period of manufacture, I designed a preliminary experiment to simulate a shot tower. Using available materials, I took some 40/60 tin/lead solder in approximately eighth-inch-diameter wire form. Using a propane torch while standing, I melted several droplets into a container of cold tap water on the floor. The drop height was approximately five feet. The lead did not have sufficient time to solidify. I therefore increased the height to seven feet by standing on a chair. The results of both tests are shown in figure 10.5. Nearly the opposite effect than expected was achieved when I increased the drop height, however:

MUSKET BALL AND SMALL SHOT IDENTIFICATION

10.5. Experimental results of dropping molten lead alloy into cold water from a height of 5' (*left*) and 7' (*right*).

10.6. Experimental shot (*left*) made by dropping molten lead ¼" from the water shown next to original shot (*right*) from the shipwreck.

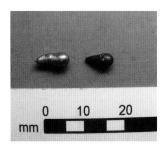

10.7. Teardrop-shaped experimental shot (*left*) shown next to actual shot (*right*) from the shipwreck.

not only did the lead not have enough free fall time to solidify, but the lead splatter was worse. I then decided to decrease the height and observe the effect.

The closer to the water, the more spherical/apple-shaped the drops became, and they began developing the concave dimple. I reduced the height to less than a quarter of an inch from the water. The resulting shot produced are shown in figure 10.6 next to actual shot recovered from the ship remains.

The results clearly show a concave dimple. All the experimental samples were weighed and found to be 0.2–0.3 grams, which is consistent with the largest quantities of bird shot found on the ship. The experimental shot were more flat, or less spherical, than the original shot. This was to be expected. The amount of heat per unit mass required to change a liquid to a solid is known as the latent heat of fusion. The amount of heat required is equal to the value of the latent heat of fusion multiplied by the change in temperature. Tin has a latent heat of fusion of 26 British Thermal Units (BTUs) per pound, whereas lead has a latent heat of fusion of 10 BTUs per pound (Rohsenow and Hartnett 1973: 3–19). This means that pure lead will take less time to cool in the water, and thus pure lead droplets using this method will be more spherical.

Further evidence that this method was used to produce the small shot found on the shipwreck was that during the experiment, the shot took a teardrop shape when the lead alloy was being melted and the heat was not applied uniformly. Figure 10.7 shows a teardrop-shaped shot produced during the experiment (left) and one found on the ship (right).

However, making shot one drop at a time is not practical. An earlier technique of producing small shot was known as the Rupert method. Prince Rupert (1619–82) of Germany was a noted soldier, admiral, scientist, sportsman,

10.8. Reproduction colander for manufacturing small lead shot using the Rupert method. *(Image courtesy of Peter Goebel, Goose Bay Workshops, Bridgeville, Delaware.)*

colonial governor, and amateur artist during the seventeenth century. In 1665 Robert Hooke published an improved method for manufacturing shot of varying sizes as communicated by Prince Rupert (Hamilton 1987: 132). Molten lead is poured through a colander, like the reproduction shown in figure 10.8, which creates molten lead droplets. The lead immediately falls into water and is cooled.

Shot made by the Rupert method were discovered on the *Queen Anne's Revenge* (Courtney Page [Queen Anne's Revenge Conservation Lab manager], personal communication, August 29, 2010), Blackbeard's flagship, which sank off the North Carolina coast in the early eighteenth century. It was a French slave ship that was captured by Blackbeard (aka Edward Teach or Thatch) in 1717. Examples of some of this shot are shown in figure 10.9.

This bird shot has the same distinctive dimple and "apple" shape as shot found on the World Trade Center ship and the experimental shot. The Rupert method can rapidly produce lead shot, making it a very practical method. The first shot tower was not conceived for a full century after the Rupert method became widespread.

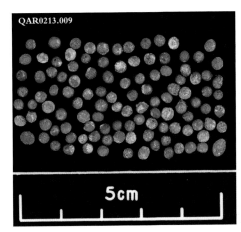

10.9. Rupert shot recovered from the *Queen Anne's Revenge*, which sank off North Carolina. *(Photograph courtesy of North Carolina Department of Cultural Resources.)*

Shot Towers

Thanks to eighteenth-century wars, especially the American War for Independence, the Industrial Revolution accelerated, making new products by faster and more mechanized means. Musket manufacturing began to become standardized based on the technology of the British Brown Bess and the French "Charleville" musket. Guns became more affordable, which created an increasing demand for ammunition, including small shot, primarily for hunting. In 1782, William Watts, a plumber from Bristol, England, was awarded a patent for making lead shot that was "solid throughout, perfectly globular in form, and *without the dimples*, scratches and imperfections, which other shot, heretofore manufactured, usually have on their surface" (emphasis added). Note the specific reference to dimples. Watts determined that when you run molten lead through a sieve and let it fall from enough height into a water tank, surface tension of the molten lead forms the droplets into almost perfect spheres, and they will harden enough to not deform when hitting the water surface (Minchinton 1990). Watts developed the concept of the shot tower, which could manufacture ammunition from musket balls to bird shot at very fast rates. This reduced the cost of the shot, making it more affordable to the average person.

Tons of lead shot were being imported from England to America until President Thomas Jefferson imposed the Embargo Act in 1807. During the early period of the Napoleonic Wars, France and Great Britain began seizing ships from neutral nations such as the United States. President Jefferson responded by banning trading with both nations (Avery 2008).

This embargo created a shortage in many products including musket shot. Because of this, in Philadelphia in 1808 John Bishop, Thomas Sparks, and James Clement erected the earliest shot tower built in America (Avery 2008). This became known as the Sparks Tower.

One shot tower that still stands as a historic site is in the center of Baltimore, Maryland (figure 10.10, left photograph). Erected in 1828, it was called the Phoenix Shot Tower. The name was later changed to the Merchants' Shot Works. The plaque on the tower states,

FOR MAKING SHOT. MOLTEN LEAD, POURED THROUGH A SIEVE AT THE TOP, DROPPED INTO A TANK OF WATER INSIDE THE BASE. HEIGHT 234 FEET, 3 INCHES: DIAMETER AT BASE 40 FEET, AT TOP 20 FEET.

This tower produced 500,000 pounds of shot per year that were packaged in 25-pound bags (figure 10.10, right photograph). As described on the edifice plaque, it was the tallest building in the United States until the Washington Monument (555 feet) was built. At its peak, it was one of the largest suppliers of

10.10. The left photograph shows the Phoenix Shot Tower, later named the Merchants' Shot Works, which still stands in the heart of Baltimore, Maryland. The right photograph shows a bag from the shot works. *(Photograph provided by the Museum of the Fur Trade Quarterly.)*

small shot in the country. It was abandoned because of rising costs and newer, more efficient methods of shot making being developed.

Bliemeister Method

Modern shot making appears to have reverted back to a process similar to the Rupert method but more mechanized and using hot water. This process eliminates the dimples in the shot. The Bliemeister method is named for Louis W. Bliemeister of Los Angeles, California, who patented it (U.S. Patent 2,978,742, dated April 11, 1961). This method is a process for making lead shot in small sizes, primarily bird shot (#6–#9), which has largely replaced the shot tower method. Antimony, added for hardness, lowers the melting point of lead (OVGuide 2013). The process is as follows:

1. Molten lead passes through successive screens that are repeatedly agitated.
2. Lead droplets fall less than one inch into hot water (as opposed to a ten-story drop).
3. The near-spherical pellets roll down an incline and are sorted by shape.

4. The nonrounded, rejected shapes are recycled and remelted. (National Institute of Justice 2013)

The water temperature controls the cooling rate of the lead, while the surface tension brings the ball into a spherical form. The shot is then coated with graphite to prevent oxidation. Variations of this method have also been used to make larger shot.

Cold Swaged Buckshot

Today, most buckshot is made from lead wire using a process known as cold swaging:

1. A strand of wire is fed between two counter-rotating wheels that meet rim to rim.
2. Hemispheres are milled into the rim of each wheel.
3. Rotation is timed so that the hemispheres in each wheel meet as the wheels turn.
4. Round balls are formed where the cavities meet and excess lead is flattened.
5. After rolling, the flattened strip is tumbled to remove the excess lead and free the pellets. The tumbling action also improves concentricity of the pellets.
6. Lead pellets may be plated with copper or nickel before loading. A tough outer coating helps protect the pellets from deformation. This step is reserved for premium-grade ammunition. (National Institute of Justice 2013)

"Pewter" Buckshot

As explained in chapter 8, lead was in short supply for the Americans during the early stages of the Revolutionary War. Some musket balls were made with an alloy of lead and tin. The tin most likely came from melting down pewter objects, which could be blended with the pure lead to stretch the amount of available lead. Apparently the same was true for buckshot. Figure 10.11 shows two buckshot found in Burlington County, New Jersey, by Bob Campbell. The color and lack of patina indicate that these shot were likely made of such an alloy. The right buckshot appears to have a higher content of tin. Both of these artifacts were found using a metal detector in an area that produced numerous eighteenth-century artifacts, including musket balls.

10.11. Buckshot made with lead/pewter alloy found in Burlington County, New Jersey.

Epilogue

This is not the end but merely the beginning of the analysis of musket balls and other early lead projectiles and artifacts. Much more work needs to be done in the areas of chemical analysis, forensic analysis of what target types cause specific deformation patterns in the lead, identifying the original sources of the lead, identifying traces of blood on oxidized bullets, and more. The object of this book is to set a basic set of standards for identifying and correctly interpreting the use of lead artifacts and stimulating the minds of future researchers in the fields of conflict archaeology. I have personally learned more than I thought possible about the simple lead sphere called a musket ball. To my wife I say, "Much more than one page and one picture!"

Appendix A

The Chronology of the Early Gun

Excerpts from "Important Dates in Gun History" (American Firearms Institute 2012), reproduced by permission. Refer to www.americanfirearms.org/gun-history/ for the complete list.

Date A.D.	Event
850	The first reference to gunpowder is probably a passage in the Zhenyuan Miaodao Yaolüe, a Taoist text tentatively dated to the mid-800s.
1000	Earliest known representation of a gun (a fire lance), Dunhuang.
1200	Gunpowder goes west with Chinese traveling in Mongol Empire.
1247	The first record of the use of gunpowder in Europe is a statement by Bishop Albertus Magnus in 1280 that it was used at the 1247 Siege of Seville.
1267	Roger Bacon gives an account of gunpowder in his *Opus Majus*. [Between 1257 and 1265, Bacon wrote a book of chemistry called *Opus Majus* which contained a recipe for gunpowder.]
1326	The earliest picture of a gun is in a manuscript dated 1326 showing a pear-shaped cannon firing an arrow. "De Nobilitatibus Sapientii Et Prudentiis Regum" by Walter de Milemete.
1346	The invention of cannon preceded by a century that of small-arms, and it was by a gradual reduction in the size of the former that the latter were produced. There is speculation about earlier use of cannon but there is evidence of their use at the battle of Cressey, in 1346.
1354	Traditional date for the German monk Berthold Schwartz to "invent" gunpowder.
1364	First recorded use of a firearm.
1375	Hand guns were known in Italy in 1397, and in England they appear to have been used as early as 1375.

APPENDIX A

1400	Hand Gonne—firearm. [The earliest "hand gonne" was developed in the fifteenth century, but was not a great influence in battle. It was a small cannon with a touch-hole for ignition. It was unsteady, required that the user prop it on a stand, brace it with one hand against his chest and use his other hand to touch a lighted match to the touch-hole. It had an effective range of only about thirty to forty yards.]
1424	The first mechanical device for firing the handgun made its appearance. Records show armor being penetrated by bullets and the handgun becoming a weapon capable of rudimentary precision.
1425	Matchlock ("arquebus") introduced. Uses a "serpentine" to arc a lit wick into the flash pan loaded with a finer grade of gunpowder. Guns were fired by holding a burning wick to a "touch hole" in the barrel igniting the powder inside. [The matchlock was a welcome improvement in the mid-fifteenth century and remained in use even into the early 1700s, when it was much cheaper to mass produce than the better classes of firearms with more sophisticated ignition systems. The matchlock secured a lighted wick in a moveable arm, which, when the trigger was depressed, was brought down against the flash pan to ignite the powder.]
1475	Invention of the arquebus or bow-gun. A spring let loose by a trigger threw the match, which was fastened to it, forward into the pan which contained the priming powder. It was from this spring that the gun took its name. [The arquebus (sometimes spelled harquebus, harkbus or hackbut; from Dutch haakbus, meaning "hook gun") was a primitive firearm used in the 15th to 17th centuries. Like its successor, the musket, it was a smoothbore firearm, although somewhat smaller than its predecessors, which made it easier to carry.
1498	Rifling was invented. [The first rifled gun barrels were made in the 1400s. This early date may be surprising, however it makes sense when one considers that arrow makers had learned to angle the fletchings on an arrow's shaft to make it spin as it flew through the air, giving it greater stability. This technique carried over to firearms.]
1509	The first wheel lock (fire-lock) or "rose lock" was invented. Some believe Leonardo da Vinci was the inventor.
1517	The Wheel Lock introduced. Uses iron pyrite rather than flint.[It is also said to have been invented by Johann Kiefuss of Nuremberg in 1517, and the idea probably came from the spring driven tinder lighter in use at the time. The idea of this mechanism is simple. A flint held by a hammer-cock was dropped onto a spinning metal wheel the friction of which showered sparks, igniting the gun

powder in a pan, which in turn ignited the powder in the barrel firing the weapon.]

1517 The snaphaunce lock, the forerunner of the flintlock, was invented.

1526 Beretta Firearms founded [sold 185 arquebus barrels to the Arsenal of Venice].

1540 Rifling was improved upon and appears [added for clarity] in firearms. Rifling refers to helix-shaped pattern of grooves (cuts) and lands (raised part of groove) that have been formed into the barrel of a firearm. It is the means by which a firearm imparts a spin to a projectile around its long axis, to gyroscopically stabilize it to improve accuracy and stability.

1612 The muzzleloading, smoothbore flintlock musket was invented as an improvement on the matchlock and wheel-lock muskets. [The flintlock was developed in France around 1612. A key contributor to this development was Marin le Bourgeoys, who was assigned to the Louvre gun shops by King Louise XIII of France.

The flintlock's manufacture slowly spread throughout Europe, and by the second half of the century it became more popular than the wheel lock and snaphaunce. The main difference between the flintlock and snaphaunce is that in the flintlock the striking surface and flashpan cover are all one piece, where in the snaphaunce they are separate mechanisms. This made the mechanism even simpler, less expensive, and more reliable than its predecessor.] The standard flintlock gun was introduced. [The flintlock solved a longstanding problem. Sometime in the late 1500s, a lid was added to the flash pan design. To expose or protect the powder, the lid had to be moved manually. The flintlock mechanism was designed to push back the lid and spark a flint at the same time. The flintlock ignition system reigned for two centuries, with virtually no alteration. A flint could be used for around 50 shots after which a new edge would be needed cut by the expert hands of a "knapper."]

1640 The bayonet was introduced by the French; it was a long narrow blade with a wooden plug handle and was simply dropped into the muzzle of the musket.

1690 The "Brown Bess" was known in Ireland as a "King's Arm" from its use by William at the Battle of the Boyne and would be used by the British Army for over 100 years.

1746 The French introduced the double-necked hammer and the steel ramrod. [The double-necked hammer or cock was not a new invention, it is found on dog locks of 1670 and other early arms.]

1763 The French introduced the muzzle band with a funnel or guide for the ramrod and acorn sight integral with the band.

APPENDIX A

1774   The Ferguson rifle, designed in 1774, was the first English breech-loading rifle made for military use. Never developed beyond an initial order of 100 rifles.

1789   The first patent for single trigger locks for double arms (James Templeman, Pat. No. 1707).

1805   The percussion cap ignition system developed and patented. [Developed by the Reverend John Forsyth of Aberdeenshire, Scotland. This firing mechanism was a great advancement from its predecessors because it does not use an exposed flash-pan to begin the ignition process. Instead, it has a simple tube, which leads straight into the gun barrel. The key to this system is the explosive cap, which is placed on top of the tube. The cap contains fulminate of mercury, a chemical compound that explodes when it is struck.]

1811   The first serious military breechloader was an American invention, Colonel John H. Hall's patent. [This was made first as a flintlock, then as percussion, and is the first breech loader officially adopted by any army. The flintlocks were made till 1832, the percussion model from 1831.]

1814   The copper percussion cap was invented.

1826   The percussion cap came into universal use on private arms.

1835   The rim fire cartridge evolved naturally out of the percussion cap and was first made by Flobert of Paris.

The American Firearms Institute gun history chronology is based on *A History of Firearms* by Major H. B. C. Pollard; "Notable Gun Dates" in Edgar Howard Penrose, *Descriptive Catalog of the Collection of Firearms in the Museum of Applied Science of Victoria [Australia]*, Museum of Applied Science of Victoria Handbook No. 1 (1949); *Firearms*, by Howard Ricketts (G. P. Putnam's Sons, 1962); Library of Congress Catalog 62–13080; *Weapons: An International Encyclopedia from 5000 BC to 2000 AD* (St. Martin's Press, 1990).

    See W. Y. Carman, *A History of Firearms from Earliest Times to 1914* (1955); A. J. Cormack, *Small Arms in Profile* (1972); E. C. Ezell, *Small Arms of the World*, 11th ed. (1977); J. Ellis, *The Social History of the Machine Gun* (1973); Graham Smith, *Gun—A Visual History*, featuring material from Civil War weapons.

Appendix B

An Analysis of Dentition Marks on Musket Balls

Henry M. Miller

At the request of Daniel Sivilich and Garry Wheeler Stone, lead projectiles recovered from the Revolutionary War battlefield of Monmouth, New Jersey, were analyzed. The purpose was to attempt to identify markings on the specimens of shot that could be attributable to human or animal mastication. Legend and preliminary inspection suggested that humans did bite into lead shot, while animals such as pigs could have attempted to chew the bullets as they foraged. This report presents the results of the study.

The Sample

Of the approximately nine hundred examples of shot recovered thus far from Monmouth Battlefield State Park through the efforts of the Battlefield Restoration and Archaeological Volunteer Organization (BRAVO), thirty-nine specimens were submitted for study that display distortions potentially caused by human or animal teeth. Subsequent study has suggested that the distortion seen on four of these is more likely because of impact damage rather than deformation due to chewing; these examples were excluded from further consideration. The remaining sample of thirty-five shot represents 3.8 percent of the total collection of musket balls recovered from the site, indicating that only a small proportion of the shot displays such distortions.

The full title of this report is "An Analysis of Marks on Musket Balls Recovered from the Revolutionary War Battlefield of Monmouth, New Jersey, and Comment on Biological Causes of Shot Deformation."

APPENDIX B

Methodology

The effort to identify impressions on the shot involved two separate activities—physical inspection/comparison and experimentation. The possibility that humans or animals bit into the artifacts and created the impressions required comparison to known surface characteristics of teeth from various species. Analysis therefore involved several steps.

First, the teeth in complete mandibles and maxilla were selected of likely species that could have chewed the shot. Beside humans, primary candidate species included swine, black bear, raccoon, opossum, and gray squirrel. Guiding this decision-making process was information about which animals are known to eat nuts, acorns, and tubers, typically foraging along the land surface. It is likely that one or more of these species could have mistaken lead shot for a nut or tuber while seeking food. For a comprehensive analysis, however, dogs and cats and grazing or browsing animals such as cattle, sheep, horses, and white-tailed deer were also included, even though these animals are much less likely to have bitten into lead bullets with any frequency. The depressions and distortions on the shot would be the negative impressions of teeth in lead, so the appearance of these impressions on a neutral background was necessary in attempting to identify them. This was accomplished by using dark blue Play-Doh, rolling out a flat segment of it and gently pressing this onto the teeth of a known animal's mandible or maxilla. Through this process, a data set of known tooth depressions identical to what one would expect to observe on the bullets was created.

Second, the bullets were compared with the Play-Doh tooth depressions from these known species. This conclusively eliminated cattle, sheep, horse, deer, and bear as possibilities. It is less likely that raccoons or opossums caused the marks, although a few of the small impressions could have been produced by one or the other of these animals. Human and swine are the most probable candidates, with one exception. This exception is a shot that displays numerous areas with shallow parallel grooves, possibly from gnawing by a squirrel, rat, or smaller rodent. Analysis indicates that premolars and molars for both humans and swine are the teeth with the most distinctive markings. Incisors and canines leave deeper puncture-type impressions that do not appear to be readily identifiable to species. In mastication, one would assume that while incisors and canines may initially grasp the ball, the chewing action would move it to the premolars and molars rather rapidly. An activity in which this would not be the case is if a soldier held the ball in his mouth while loading the weapon, thereby imparting incisor and/or canine impressions on two sides of the ball.

Next, it was necessary to compare the bullet impressions with actual teeth. For humans, it was assumed that all individuals would have adult dentition. For swine, this is complicated because the deciduous teeth have a different pattern than adult teeth, and heavy wear will eliminate the distinctive cusp patterns over time. To overcome this difficulty, it was necessary to assemble a wide range of swine mandibles and maxilla that reflect both age and wear differences. For this purpose, various archaeologically recovered swine jaws from colonial sites were pulled from collections, along with modern specimens, thereby providing a good data set of how swine teeth change through wear. The markings on each bullet were then inspected visually and directly compared with the actual dentition. Identification was accomplished by determining whether the deformation pattern on the shot was visually similar or identical to that displayed by the teeth. Wherever possible, a direct physical match of the shot depressions with actual teeth was attempted. Variation between individual pigs makes a precise match unlikely, but swine teeth do have a strong genetic-based regularity in form, and they wear in a generally predictable manner. Most of the archaeological specimens were probably from free-ranging swine, not penned animals. Given the wide spatial distances between the recovery locations of the Monmouth shot, it is likely that any swine-related marks on these specimens would have also been produced by foraging animals, not pigs confined to a pen. It is reasonable to expect that a good correspondence between some teeth and some lead shot impressions will be found, and this was the case for the Monmouth collection.

One major difficulty made obvious by this study is that animals or humans can chew a lead ball multiple times, and each subsequent bite may distort or obliterate the previous teeth impressions. Hundreds of marks were observed on these bullets, and they include shallow and deep depressions, shallow scrapes, flattened sections with ripples and distortions, and areas with irregular surfaces. Only a few displayed sufficient detail to be positively identified as teeth marks from a specific species. Also, all of the artifacts are covered with a corrosion layer, which partially conceals or obscures some of the marks.

Experimental Evidence

The second part of the analysis was experimenting with modern shot to understand the likely sources of the marks. Dan Sivilich had lead musket balls cast for this purpose. (Note that in chapter 7, Sivilich reports he later found that these were cast from a lead alloy, not pure lead, potentially making them harder than pure lead.) These were then bitten and chewed by Sivilich and myself to duplicate human tooth depressions. This provided direct evidence regarding the

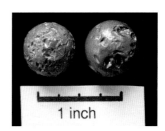

B.1. Modern musket balls with human chewing marks. *(Photograph by Henry Miller.)*

types of depressions produced by chewing (figure B.1). We found that it was difficult for humans to make deep impressions. The lead balls were hard and human musculature can only create relatively shallow dents and scratches in the surface of large balls. Over time, however, it is possible to cover the surface of a bullet with these shallow depressions and irregularities, creating a distinctive rough appearance. Smaller balls, such as buckshot, have less mass and seem to be easier for human mastication to deform. There is definitely a thickness/mass factor involved in the degree to which the human jaw can modify a lead shot.

In a continuation of this experiment, other balls were given to an adult Ossabaw pig at the Tobacco Plantation exhibit at Historic St. Mary's City. These bullets were embedded in apples and fed to the mature pig at the exhibit. The pig chewed the first apple and bullet for a brief time and then dropped the bullet from its mouth. With the second apple, the pig bit into it once and immediately spat out the bullet, leaving only two depression points on the ball. Clearly, this pig learned not to chew on lead after a single experience, providing some insight regarding the intelligence of the species. Examination of the first, more chewed specimen revealed some deep marks on the surface, some of which have a tooth pattern. While some of these are similar to the shallow depressions produced by humans, other depressions and scrapes are deeper and larger. One should note that this was an old pig (ten years old) with worn dentition; younger swine will probably chew more aggressively and produce greater deformation.

Pope's Fort Shot

To better evaluate the possibility that humans actually chewed shot as a common activity, a sample of shot from an area not likely to have been accessible to pigs or other animals was needed. The available archaeological sample used for this component of the analysis is bullets recovered from Pope's Fort at St. Mary's City. This fort dates to the 1645–circa 1655 period and is directly linked to the English Civil War. Pope's Fort surrounded the home of Maryland's first governor, Leonard Calvert, and this vicinity was part of the yards of the Calvert house during the seventeenth century. We can assume, from the multiple generations of fences found there by archaeologists, that pigs were generally excluded from the yards. Historically, we also know that this structure was a home to two governors and served as an official statehouse for Maryland until 1676. In a further refinement of the sample, only specimens deeply buried in the fill of the fort moat since circa 1650 were examined. While this sample is much earlier than the Monmouth bullets, it does provide a sample of colonial shot in a good, undisturbed archaeological context. Most of the specimens are smaller than the Monmouth shot, being buckshot- or "goose" shot–sized artifacts, although there is a range of sizes represented.

Study of these lead balls from Pope's Fort revealed bullets in three conditions. Most are unaltered and round in shape and display smooth surfaces; they are little changed from the time they emerged from the shot mold, aside from a little corrosion. These are unused shot, either lost or intentionally discarded. The second category is represented by a few examples that are significantly deformed from having been fired and impacting on a hard material. The shot in the third group retain a circular shape but have rough irregular surfaces with numerous shallow depressions and scrapes. Figure B.2 shows the unaltered shot and two examples of the chewed shot recovered from the fill layers in the moat of Pope's Fort.

These roughened examples have surfaces that are virtually identical in appearance to the modern musket balls chewed by Sivilich and myself. Significantly, none of these specimens shows deep impressions or is severely distorted. These Pope's Fort balls with surfaces completely covered with tooth impressions appear to have been chewed by humans for a considerable time. This did not greatly distort the round shape of the ball, however, for only the surface is heavily modified. It is important to note that some of these chewed shot are found in the same layers as the unused specimens, clearly indicating that these irregular surfaces were not caused by the depositional environment. This suggests that the practice of chewing lead shot, perhaps to aid in promoting salivation when thirsty, dates back to the early years of colonial settlement in America.

B.2. Shot recovered from Pope's Fort: *top row*, human chew marks; *bottom row*, unaltered shot (18ST1-13-1639 J and T). *(Photograph by Henry Miller.)*

Results

Of the thirty-nine specimens from Monmouth Battlefield submitted for analysis, four (specimens 9K15-3, 90M23RF2, 91C16DS4, and 91A6DS-1) are impacted shot and were excluded from further study. One of these (artifact 9K15-3) shows a flat rectangular impact area, the cause of which is uncertain. This leaves thirty-five musket balls with potentially identifiable impressions. Of these, seventeen are quite distorted, indicating considerable jaw strength. Some of them have an appearance resembling used chewing gum. The depth of these depressions and the amount of distortion is more than can be typically produced by human jaw muscles. However, no positive teeth marks could be identified on these seventeen specimens, in part because of multiple instances of overlapping biting. They are probably swine-chewed, but this cannot be positively confirmed. These specimens are listed in table B.1.

APPENDIX B

Table B.1. Monmouth Battlefield shot with unidentified teeth-related distortion

Artifact	Description
Belle Terre DS-1	Heavily distorted, chewed nearly flat
Belle Terre DS-2	Possibly impacted and later chewed
90E26-2CR	Small ball with deep teeth marks
91C16RPG	Numerous teeth impressions
92E2TW1	Heavily chewed
92M19TW1	Many marks, too deep for human
94E1MH1	Marks too deep for human
94E14JM06	Round ball with teeth marks, some deep
206-3-208	Largely flattened from chewing
207-3-506	Flattened from chewing
224-2-965A	Numerous teeth marks
224-4-439	Rounded with some deep impressions
224-4-808	Round but with some deep chew marks
228-4-585	Mostly round with some deep marks
234-2-1006	Round with a few shallow marks, one very deep impression
234-3-673A	Chewed flat
234-9-804	Many impurities, chew marks over surface (if the impurities made the lead softer, these could be human marks)

B.3. Artifact 93L13DS-4 with swine teeth impressions. *(Photograph by Henry Miller.)*

B.4. Artifact 224-4-809 with a shear mark probably caused by swine. *(Photograph by Henry Miller.)*

B.5. Artifact 227-2-778 with swine molar impressions. *(Photograph by Henry Miller.)*

Likely Swine-Modified Bullets

There are eleven specimens with identified animal impressions. Of these, ten have marks that appear to have been created by swine. All the bullets have numerous marks, but only a few of these were potentially recognizable. Ultimately, single impressions on each of the ten specimens could be identified. Because of their significance, these are discussed below.

Rodent Modification

Another bullet (90M16RP4) is unique in this sample. Its surface displays numerous short parallel grooves that are very shallow. These occur in multiple directions over the surface of the bullet and vary in length (figure B.7). Some groups of these marks consist of multiple parallel grooves. One hypothesis is that they were caused by forceps used to extract the bullet. However, there should be numerous matching marks on opposite sides of the ball if the marks were caused by forceps grasping it, and I could not find clear matches.

To acquire more information, I conducted an experiment with modern forceps on a newly cast lead shot. This produced matching depressions on opposite sides of the ball. In addition, the ends of the short parallel lines formed by the grooves in the gripping surface of the tool all stopped in a straight line, representing the edge of the instrument. These features were not seen on the bullet. (A pair of Revolutionary War period forceps was not available for comparison,

Table B.2. Specimens from Monmouth Battlefield likely mauled by swine leaving distinct teeth marks

Artifact	Description
90M9RP-15	Good match for a left mandibular first molar from a young swine, the mark is from the rear portion of the tooth, and there is no indication of an adjacent M2 impression, suggesting that the tooth had not yet erupted
90M16DS-3	Shows the impression of what might be a maxillary P4 swine premolar with some wear on it
91D20DS-5	Possible match for a right mandibular first molar from a young pig; the M1 has only moderate wear, so the other teeth would have been deciduous
93L13DS-4	Possible molar cusp impression from a mature swine, most likely a second or third molar; the impression is deep
224-2-80	Match to P3 and P4 premolars from a left maxilla; part of each tooth appears on the bullet, with a good match to a complete skull of a mature swine
224-2-900	Pair of indentations that may fit a swine deciduous right mandibular P3–P4 premolar junction, with the joint between them seen in the lead
224-4-809	One long crushing scar, the striations and shape of which match the interior (lingual side) edge of a maxillary second molar that was worn relatively flat, indicating an older swine (it matches very closely teeth in a skull section found at the Webbs Landing site in Delaware [70-33]) (figure B.4)
227-2-778	Clear marks of a well-worn molar with the small indention of the dentin surrounded by the harder enamel (might be a first or second mandibular molar but clearly is a swine tooth); lead artifact is flat with tooth marks on both surfaces, indicating intense crushing action (figure B.5)
227-3-83	Repeated series of chew marks, like a sequence of individual bites; one set of these matches the cheek (buccal) side tooth pattern of the maxillary left fourth premolar and first molar junction of a mature swine (figure B.6)
234-1-467	Appears to be the pattern of a slightly worn molar, with the swine fissure pattern on the tooth displayed distinctly

B.6. Artifact 227-3-83 with swine bite marks. *(Photograph by Henry Miller.)*

B.7. Artifact 90M16RP4 with rodent gnaw marks. *(Photograph by Henry Miller.)*

but similar results would presumably be obtained. Further investigation would be of value.)

Many of these depressions are not parallel but consist of clusters of double grooves running at slight angles to each other. The occurrence of these marks on all portions of the ball, their multiple directions, and the appearance of isolated double-groove impressions suggest another cause: such striations are very similar to rodent gnaw marks observed on animal bone from archaeological sites. If the impressions are attributable to rodent activity, then the grooves are caused by the incisors of a small rat or squirrel or a mouse. Comparison indicates that the marks are smaller than the incisor impressions that would be made by a gray squirrel or a European rat. This interpretation raises the question of why a rodent would chew on a musket ball. There is some evidence to suggest that lead tastes sweet to these animals. At times, this gnawing by a rodent was so extensive that it significantly misshaped the ball. Since writing the original version of this

APPENDIX B

study, I have examined another specimen, this one from the 1780 battlefield at Camden, South Carolina. As figure B.8 shows, as much as one-half of this ball was eaten away, with incisor marks from varied directions. The side view (figure B.9) shows the remains of this round shot with a series of dual incisor marks at the edge, made as the animal sought to cut into the lead. A rodent of some type, probably a rat or squirrel based on the incisor size, consumed a considerable part of this shot.

Likely Human-Modified Bullets

These musket balls are randomly covered with shallow depressions and nicks. Experimentation suggests that these are mostly likely the marks made by human teeth. They are not deep and tend to be more puncture shaped, as human canines and premolars would produce. However, no specific human molar or premolar pattern was identified on these specimens. Table B.3 lists these specimens, and figure B.10 displays some specimens from Monmouth.

In addition, specimen 234-9-804 listed in table B.1 could have marks made by a human. It is covered with depressions and dents. Some are deeper than seen on the above specimens or produced by experimentation with human teeth. However, this shot has numerous impurities mixed into the lead from casting; it has more impurities than any of the other balls examined in this study. This could have made it softer and allowed greater deformation to occur by human action.

What is intriguing about these musket balls with likely human teeth marks is their archaeological distribution. Four specimens (92M19TW-4, 224-1-342, 224-2-639, and 224-3-103) were all found at the Parsonage Farm site. In addition, the possible human-chewed specimen (224-9-803) listed in table B.1 is from the Parsonage Farm area. Thus, five of the seven balls with possible human teeth marks were found in that location. This concentration of likely human-chewed balls is notable and could be a significant clue. Does it indicate that troops were assembled there for a time and chewed on lead balls to relieve their

B.8. Top view of a musket ball from Camden, South Carolina, with major rodent chewing evident from multiple directions. *(Photograph by Henry Miller.)*

B.9. Side view of the Camden, South Carolina, shot showing rodent teeth marks. *(Photograph by Henry Miller.)*

Table B.3. Specimens from Monmouth Battlefield with likely human teeth marks

Artifact	Description
92M19TW-4	Surface displaying shallow marks
206-3-37	Could be human but not as certain; shallow with a few deep pits
224-1-342	Surface covered with numerous shallow marks
224-2-639	Shallow marks over most of surface
224-2-986	Dropped or possibly impacted ball but with shallow tooth marks on the nonimpacted surface
224-3-103	Surface with numerous shallow teeth marks

thirst? Or did some other activity occur there that would have involved biting on lead balls? Historical documents about the battle should be investigated so that these possibilities can be better evaluated: archaeology is indicating that something specific occurred in that area that involved people chewing musket balls.

Summary

Analysis of the shot recovered from Monmouth Battlefield indicates that a small percentage of these artifacts display marks caused by chewing. Study of these marks and comparison with known specimens suggest that both humans and animals were responsible for the marks on these bullets. Most of the identified marks seem to have been caused by swine. This is likely explained by postbattle foraging by swine, either by free-ranging animals or by pigs being allowed to glean in the fields after harvest. A round shot would have initially seemed to be an acorn or nut to the pig. One possible rodent-gnawed example is also present in the Monmouth collection, and rodent deformation has been identified on shot from other contexts.

The ability of swine to chew and crush shot is clearly demonstrated by a seventeenth-century swine mandible fragment recovered from a cellar at the Thomas Pettus site near Jamestown (44JC33). As figure B.11 shows, the third molar has lead embedded in its fissures, the result of the pig chewing an object that was almost certainly a lead shot. Heavy deformation of large-caliber shot is most likely the result of swine.

B.10. Artifacts with likely human bite marks: specimens 92M19TW-4, 224-3-103, and 224-2-639. *(Photograph by Henry Miller.)*

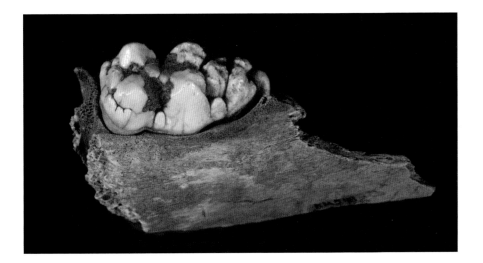

B.11. Seventeenth-century swine mandible fragment (44JC33) from the Pettus plantation site, Virginia, with lead embedded in the molar fissures from chewing a piece of lead. *(Photograph by Henry Miller.)*

APPENDIX B

However, identifying a likely biological cause that produced the malformation of a shot demands careful analysis of the specimen. Deformation can be of animal or human origin. One historical account by a surgeon with the Continental army during the American Revolution, James Thacher, provides an important detail:

> [T]he culprit is sentenced to receive one hundred lashes, or more.... The culprit being securely tied to a tree, or post, receives on his naked back the number of lashes assigned him, by a whip formed of several small knotted cords, which sometimes cut through the skin at every stroke. However strange it may appear, a soldier will often receive the severest stripes without uttering a groan, or once shrinking form [*sic*, from] the lash, even while the blood flows freely from his lacerated wounds. This must be ascribed to stubbornness or pride. They have however, adopted a method which they say mitigates the anguish in some measure, it is by putting between the teeth a leaden bullet, on which they chew while under the lash, *till it is made quite flat and jagged.* (Thacher 1823: 222–23 [emphasis added])

Under extreme stress, it is possible for a piece of shot to be heavily crushed by a person, as Thacher apparently observed. But he does not describe the size of the leaden bullet they used. Smaller-caliber balls like buckshot likely could be flattened in such a situation. But a large ball, such as one 0.69 inches in diameter, would be far more difficult to significantly deform by human teeth, as our experiments with modern shot indicate. Some smaller pieces of lead that are flat and jagged could indeed be the result of human mastication, while larger specimens with marked alteration are more likely attributed to the more powerful jaws of swine. Obviously, if clear teeth marks can be identified, a definitive conclusion can be reached. In analysis, one must consider the context of recovery, the size of the specimen, and the degree of malformation. Was it in a location to which animals such as swine and rodents may have had access at some point in the past? What was its original size (probably determinable only by weight for many specimens)? And is the deformation extreme, moderate, or slight? All must be taken into account when trying to offer an explanation for a piece of chewed lead recovered from an archaeological site.

By examining the Monmouth shot as a group, I was able to define two general categories: those with heavy distortion and deep impressions and those with surfaces that display shallow depressions but little overall distortion to their basic round shape. Experimentation and size analysis suggest that the more heavily distorted group is swine-altered bullets while those with shallow depressions are human-altered. In addition, comparative data from a seventeenth-century fortification support the interpretation that the balls covered with shallow depressions and with uniformly roughened surfaces are the products of humans modifying them by chewing. "Biting the bullet" by a man or woman, however, should be considered only one of the possible explanations.

Acknowledgments

I wish to thank Garry Wheeler Stone and Daniel Sivilich for inspiring and providing the data to conduct this study. Also, gratitude is owed to George Beall of Matthews, North Carolina, for the kind loan of specimens for analysis.

Appendix C

Classifications of Small Lead Shot

A musket ball is designed to be loaded tightly into the barrel of a gun and fired as a single projectile. It must be accurately aimed to hit a target. Small lead shot is designed to be loaded loosely in the barrel of a gun and in quantity. When the load leaves the muzzle of the gun, it fans out into a cone, or "scatters," yielding a greater possibility of hitting a target. This is useful for hitting moving targets such as a running deer or ducks in flight. Thus, names such as "buck" shot, "goose" shot, and "quail" shot came into use. Rats were common on early ships. How does one shoot rats on a wooden ship without causing damage to the vessel? Very fine lead shot was used, and it became known as "rat" shot. However, these names were somewhat arbitrary for specific sizes of lead shot. So how does one distinguish small musket/pistol balls from buckshot? The following table shows the contemporary descriptions and sizes for small shot. In general terms, lead shot with diameters of 0.37 inches and above are musket balls.

Table C.1. Small shot sizes and descriptions

Number of shot per pound	Size/description	Dia. (in.)	Dia. (mm)	Weight (g)
Buckshot				
100	000	0.360	9.144	4.5359
130	00	0.330	8.382	3.4892
143	0	0.320	8.128	3.1720
173	1	0.300	7.620	2.6219
232	2	0.270	6.858	1.9551
299	3	0.250	6.350	1.5170
338	4	0.240	6.096	1.3420
Bird shot				
400	FF	0.230	5.842	1.1340
464	F	0.220	5.588	0.9776
560	TT	0.210	5.334	0.8100
672	T	0.200	5.080	0.6750
800	BBB	0.190	4.826	0.5670

Table C.1. (*Continued*)

Number of shot per pound	Size/description	Dia. (in.)	Dia. (mm)	Weight (g)
928	BB	0.180	4.572	0.4888
1,040	B	0.175	4.445	0.4361
1,136	1	0.160	4.064	0.3993
1,376	2	0.150	3.810	0.3296
1,696	3	0.140	3.556	0.2674
2,112	4	0.130	3.302	0.2148
2,688	5	0.120	3.048	0.1687
3,488	6	0.110	2.794	0.1300
4,656	7	0.095	2.413	0.0974
6,384	8	0.090	2.286	0.0711
9,088	9	0.080	2.032	0.0499
13,568	10	0.070	1.778	0.0334
21,536	11	0.060	1.524	0.0211
37,316	12	0.050	1.270	0.0122
Dust or rat shot				
73,040	Dust	0.040	1.016	0.0062
172,616	Fine dust	0.030	0.762	0.0026
1,380,925	Extra-fine dust	0.015	0.381	0.0003

Sources: Potter and Hanson 2001; ANSI/SAAMI Z299.2.

Appendix D

Bravo

BRAVO (Battlefield Restoration and Archaeological Volunteer Organization) is a federal 501(c)(3) nonprofit, volunteer organization dedicated to locating and preserving important archaeological sites. It was founded in 1999 as an organization of people from varied backgrounds with interests in history and archaeology. Since then, it has become known worldwide as a leader in developing and using new methodologies for battlefield archaeology. BRAVO members have volunteered tens of thousands of hours of their personal time working on archaeological projects and locating the hundreds of artifacts shown in this book, as well as many others.

Significant historic sites are being lost to development and looting. Reality television shows about metal detecting fail to instruct the public about responsible archaeological procedures and documentation. The result is an increase in hobbyists destroying important archaeological sites through a lack of knowledge or lack of caring for preserving our national heritage. Internet sales have inflated the values of artifacts and have made it easy to profit from artifacts obtained illegally. BRAVO has been working to educate hobbyists about their responsibilities and to bring the avocational and professional archaeological communities together.

The archaeological research being done by BRAVO and professional archaeologists helps identify historically significant sites and provides the rationale for acquisition and preservation. BRAVO strives to achieve the following:

- Increase public awareness of historic sites through archaeology and interpretation
- Assist in the enhancement, preservation, and development of these sites for public use and enjoyment
- Assist in interpretation, public education, and research connected with historical events
- Increase awareness of responsible metal detecting

BRAVO is known worldwide for its work in developing the area of electronic archaeology using metal detectors, a total station laser transit, GPS, ground-penetrating radar (GPR), and layered site mapping using geographic information systems (GIS).

Our archaeological projects focus on New Jersey state parks, with a primary emphasis on Monmouth Battlefield State Park. However, BRAVO is not restricted solely to New Jersey state parks. Other state parks, national parks, privately owned historic sites, local municipal governments, and professional archaeological consulting companies have used the services of BRAVO volunteers. Here is a list of some of the historic sites (and in the left column, their dates) that BRAVO members have worked on:

Aug. 15–16, 1689	Battle of Zboriv, Ukraine
1754–63	Fort Edward, N.Y.
Dec. 8, 1776	Battle of Two Bridges, Branchburg, N.J.
1776–	Raritan Landing British/American encampment, Piscataway, N.J.
1776–77	Continental Arms Factory at Pickering Creek, East Pikeland, Pa.
1777	Fort Montgomery, N.Y.
Jan. 3, 1777	Battle of Princeton, Princeton, N.J.
March 1777	American barracks burned by the British, Peekskill, N.Y.
June 26, 1777	Battle of Short Hills and Oak Tree Pond, Edison, N.J.
July 7, 1777	Battle of Hubbarton, Hubbarton, Vt.
Sept. 3, 1777	Battle of Cooch's Bridge, Newark and Glasgow, Del.
Sept. 11, 1777	Battle of Brandywine, Birmingham Township, Pa.
Sept. 11, 1777	Sandy Hollow section of the Battle of Brandywine, Pa.
Sept. 21, 1777	Paoli Battlefield/massacre site, Malvern, Pa.
1777	Wayne's Brigade encampment site, Washington Valley Park, N.J.
1777	East Pikeland, Pa., munitions factory and encampment at Snyder's Mill destroyed by the British
1777–78	Valley Forge encampment, Valley Forge and Lower Providence Twp., Pa.
June 28, 1778	Battle of Monmouth, Freehold/Manalapan, N.J.
1778–	Revolutionary War artillery road between West Point and New Windsor, Cornwall, N.Y.
1778–	Redoubts 3 and 4, U.S. Military Academy, West Point, N.Y.
1778–79	American Revolutionary War artillery school and encampment, Pluckemin, N.J.

D.1. Members of BRAVO at the Washington Memorial Chapel site in Valley Forge, Pennsylvania.

D.2. Members of BRAVO at the Blue Licks Battlefield State Park and Recreation Area, Kentucky. Adrian Mandzy is second from the right, followed by Joyce Bender and Zeb Weese.

D.3. Members of BRAVO at one of the Cooch's Bridge sites in Newark, Delaware, with Wade Catts (*second from the right*).

BRAVO

D.4. Members of BRAVO at Monmouth Battlefield State Park with Douglas Scott, fourth from the right.

D.5. Members of BRAVO with Douglas Scott (*fourth from the right*) and Garry Wheeler Stone (*fifth from the right*). The author is to the left of Garry Wheeler Stone.

1779–80	Jockey Hollow encampment at Saint Mary's Abbey, Morristown, N.J.
June 7, 1780	Battle of Connecticut Farms, Union, N.J.
June 23, 1780	Battle of Springfield, Springfield, N.J.
Aug. 19, 1782	Battle of Blue Licks, Mount Olivet, Ky.
Aug. 31, 1814	Battle of Caulk's Field, Chestertown, Md.
1862–64	Camp Vredenburg, Manalapan, N.J.
1864	Battle at the Crater, Petersburg, Va.
1933–42	Bass River State Park CCC camp site, New Gretna, N.J.
1933–42	Voorhees State Forest CCC camp site, Glen Gardner, N.J.

There are many faces of BRAVO. These are just a few photos of the many members of BRAVO who have contributed to this book and the knowledge of early American conflict archaeology. Without them, this book would not have been possible.

References

Adelberg, Michael S.
2010 *The American Revolution in Monmouth County: The Theatre of Spoil and Destruction.* Charleston, S.C.: History Press.

Adkin, Mark
2001 *The Waterloo Companion.* Mechanicsburg, Pa.: Stackpole Books.

Ahearn, Bill
2005 *Muskets of the Revolution and the French and Indian Wars.* Woonsocket, R.I.: Andrew Mowbray Publishers.

American Firearms Institute
2012 "Important Dates in Gun History." www.americanfirearms.org/gun-history/ (accessed November 14, 2012).

ANSI/SAAMI (American National Standards Institute / Sporting Arms and Manufacturers' Institute)
2015 "American National Standard Voluntary Industry Performance Standards for Pressure and Velocity of Shotshell Ammunition for the Use of Commercial Manufacturers." ANSI/SAAMI Z299.2-Shotshell-2015, pp. 36–37.

Avery, Elroy McKendree
1918 *Biography: A History of Cleveland and Its Environs, the Heart of New Connecticut.* Vol. 2. Chicago: Lewis Publishing.

Avery, Ron
2008 "From Musket Balls to Basketballs—The Sparks Shot Tower." *The Philly History Blog.* January 25. www.phillyhistory.org/blog/index.php/2008/01/from-musket-balls-to-basketballs-the-sparks-shot-tower/.

Babits, Lawrence E.
1998 *A Devil of a Whipping: The Battle of Cowpens.* Chapel Hill: University of North Carolina Press.

References

Bailey, Dewitt
2002 *British Military Flintlock Rifles, 1740–1840.* Lincoln, R.I.: Andrew Mowbray Publishers.

Basset, Andre
177– *La destruction de la statue royale a Nouvelle Yorck / Die Zerstorung der Koniglichen Bild Saule zu Neu Yorck.* Paris: Chez Basset.

Bell, Andrew
1778 Diary from Philadelphia during "the March of the British Army across the Jersies [*sic*]—and the Battle of Monmouth." Manuscript Group 45. New Jersey Historical Society, Newark, N.J.

Bell, J. L.
2012 "Both Poisoned and Chewed the Musket Balls." *Boston 1775.* June 18. www.boston1775.blogspot.com/search/label/John Waller.

Bilby, Joseph G., and Katherine Bilby Jenkins
2010 *Monmouth Court House: The Battle That Made the American Army.* Yardley, Pa.: Westholme Publishing.

Bretscher, Ulrich
2009 "The Two Oldest Handguns of Switzerland." *Ulrich Bretscher's Black Powder Page.* Last updated September 2009. www.musketeer.ch/blackpowder/freienstein.html.

Britannica
2014 "Bow and Arrow." *Britannica Online Encyclopedia.* Last updated February 2014. www.britannica.com/EBchecked/topic/76056/bow-and-arrow.

Burns, William E.
2005 *Science and Technology in Colonial America.* Westport, Conn.: Greenwood Press.

Calver, William Louis, and Reginald Pelham Bolton
1950 *History Written with Pick and Shovel: Military Buttons, Belt-Plates, Badges and Other Relics Excavated from Colonial, Revolutionary, and War of 1812 Camp Sites.* New York: New-York Historical Society.

Campillo, Xavier-Rubio
2008 "An Archaeological Study of Talamanca Battlefield." *Journal of Conflict Archaeology* 4 (1–2): 23–38.

Capitaine, Michel du Chesnoy
1778 *Plan de l'affaire de Montmouth, ou le Gl. Washington commandons l'Armée Americane et le Gl. Clinton l'Armée Angloise, le 28 Juin 1778.* McDougal Papers, New-York Historical Society, New York.

Caruana, Adrian B.
1979 *British Artillery Ammunition, 1780.* Bloomfield, Ont.: Museum Restoration Services.
1990 "Tin Case-Shot in the 18th Century." *Arms Collecting* 28 (1). Text available at http://www.militaryheritage.com/caseshot.htm.

Clinton, Sir Henry
1954 *The American Rebellion: The British Commander-in-Chief's Narrative of His Campaigns, 1775–1782.* New Haven, Conn.: Yale University Press.

Craig, Bruce D., and David S. Anderson
2002 *Handbook of Corrosion Data.* 2nd ed. Materials Park, Ohio: ASM International.

CTSSAR
1998 "King George's Head." *The Connecticut Society of the Sons of the American Revolution.* Originally published in *The SAR Magazine* (Winter 1998). Online document, www.connecticutsar.org/king-georges-head/.

Dodsley, J.
1778 *The Annual Register, or a View of the History, Politics and Literature, for the Year 1777.* London: privately printed.

Dolleczek, Anton
1896 *Monographie der k.u.k. österr.-ung. blanken und Handfeuerwaffen.* Graz, Austria: Akademische Druck-u. Reprint, 1970.

Drury, Robert, and Alexis Marie de Rochon
1969 *Madagascar; or, Robert Drury's Journal during 15 Years' Captivity on That Island.* New York: Negro University Press.

Fisher, Charles, Gregory Smith, Lois Feister, Nancy Davis, Christina Reith, Jennifer Bollen, Beth Horton, J. Scott Cardinal, and Lihua Whelan
2004 *"The Most Advantageous Situation in the Highlands": An Archaeological Study of Fort Montgomery State Historic Site.* Edited by Charles Fisher. Albany: New York State Museum.

Foard, Glenn
2005 "The Battle of Edgehill: History from the Field." *Battlefields Annual Review,* pp. 43–54.
2009 "English Battlefields, 991–1685: A Review of Problems and Potentials." In *Fields of Conflict, Battlefield Archaeology from the Roman Empire to the Korean War,* edited by Douglas D. Scott, Lawrence Babits, and Charles Haecker, 84–101. Dulles, Va.: Potomac Books.
2012 *Battlefield Archaeology of the English Civil War.* Oxford, U.K.: Archaeopress.

REFERENCES

Gadalla, Moustafa
2007   *Ancient Egyptian Culture Revealed.* 1st ed. Greensboro, N.C.: Tehuti Research Foundation.

Gates, Horatio
1776   Ebenezer Hazard to Horatio Gates. Letter dated July 12, 1776. Gates Papers, New York Historical Society, New York.

Goldstein, Eric, and Stuart Mowbray
2010   *The Brown Bess: An Identification Guide and Illustrated Study of Britain's Most Famous Musket.* Woonsocket, R.I.: Mowbray Publishing.

Hamilton, T. M.
1987   *Colonial Frontier Guns.* Union City, Tenn.: Pioneer Press.

Hanger, George
1814   *Colonel George Hanger to All Sportsmen.* London: privately printed.

Hartgen Archeological Associates
2008   *Archeological Investigation of the Sackets Harbor War of 1812 Battleground Village of Sackets Harbor, Jefferson County, New York.* Albany, N.Y.: Hartgen Archeological Associates.

Historic St. Mary's City
2013   *Historic St. Mary's City History.* Available at http://www.hsmcdigshistory.org/research/history/ (site updated 2013).

Hogg, Ian V.
1981   *An Illustrated History of Firearms.* New York: A and W Publishers.

Hopkins, Alfred F.
1940   "Equipment of the Soldier during the American Revolution." *Regional Review* (National Park Service, Region One, Richmond, Va.) 4 (3): 19–22. Electronic republication, http://npshistory.com/series/popular/2/ps2-1.htm.

Hornady Manufacturing Company
2010   *Hornady 2010 Catalog.* Grand Island, Neb.: Hornady Manufacturing Company.

Janská, Eva
1963   "Archaeological Investigation of the Castle of Sion." *Archeologické rozhledy* 15.

Kinkor, Kenneth J., Sharon Simpson, and Barry Clifford
2007   *Real Pirates: The Untold Story of the* Whydah *from Slave Ship to Pirate Ship.* Washington, D.C.: National Geographic Society.

Knarrström, Bo
2006   *Slagfaltet*. Saltsjo-Duvnas, Sweden: Efron and Dotter.

Koenig, Arman
2012   *Reproduction Matchlock and Wheelock Muskets*. Available at www.engerisser.de/Bewaffnung/weapons/Matchlockmusket.html (accessed December 15, 2012).

Levine, Noah
2012   "Biting the Bullet." *Dental Products Report*. February 14. www.dentalproductsreport.com/dental/article/biting-bullet.

Mackenzie, John
2013   "The Battle of Princeton." *Britishbattles.com*. www.britishbattles.com/battle-princeton.htm (accessed April 10, 2013).

Mandzy, Adrian
2012   "Using Munitions and Unit Frontage: New Evidence about the Russian Main Battle Line at Poltava (1709)." In *Fasciculi Archaeologiae Historicae: Recent Research into Medieval and Post Medieval Firearms and Artillery*, edited by Jerzy Maik, 67–76. Lodz, Poland: Institute of Archaeology and Ethnology of Polish Academy of Sciences.

Mandzy, Adrian, Daniel Sivilich, Erik Hale, Joe Marin, and Stephen McBride
2008   "A Phase I Survey of the Blue Licks Battlefield." Report on file with Kentucky Department of Parks, Frankfort.

Martin, Joseph Plumb
1988   *Private Yankee Doodle*. New York: Eastern Acorn Press.

Mayer, John
1990   "Sparks Shot Tower, 1808, 129–131 Carpenter Street, Philadelphia, PA, 19147." *Workshop of the World*. Last updated May 2007 by Joel Spivak. www.workshopoftheworld.com/south_phila/sparks.html.

McConnell, David
1988   *British Smooth-Bore Artillery: A Technological Study*. Ottawa, Ont.: Research Publications, Environment Canada—Parks.

Miller, Henry M.
2004   An Analysis of Marks on Musket Balls Recovered from the Revolutionary War Battlefield of Monmouth, New Jersey. Historic St. Mary's City, St. Mary's City, Maryland.

Minchinton, Walter
1990   "The Shot Tower." *American Heritage* 6 (1). Reprinted as *The Shot Peener* 7 (3): 22–24.

REFERENCES

Muller, John
1977 *A Treatise of Artillery, 1780.* Bloomfield, Ont.: Museum Restoration Services.

National Institute of Justice
2013 "Projectiles: Current Manufacture." In "Small Arms Ammunition," module 5 of *Firearm Examiner Training.* http://projects.nfstc.org/firearms/module05/fir_m05_t05_04.htm (accessed May 14, 2013).

Neumann, George C.
1967 *The History of Weapons of the American Revolution.* New York: Bonanza Books.

Neumann, George C., and Frank J. Kravic
1989 *Collector's Illustrated Encyclopedia of the American Revolution.* Texarkana, Tex.: Rebel Publishing.

New Jersey Historical Society
1942 *The Letters of William Alexander, Lord Sterling.* Proceedings of the New Jersey Historical Society 60, 174.

OVGuide
2013 *Watch Bliemeister Method Video.* www.ovguide.com/bliemeister-method-9202a8c04000641f80000000009bf240 and view "Lead Shot Making" (accessed May 14, 2013).

PaperMate
2013 "History of Pencils." *About Berol.* www.berol.co.uk/historyofpencils.html (accessed October 27, 2013).

Peterson, Harold L.
1968 *The Book of the Continental Soldier.* Harrisburg, Pa.: Stackpole.

Potter, Gail Debuse, and James A. Hanson
2001 "Shot Sizes—What Do They Mean." *Museum of the Fur Trade Quarterly,* 7–13.

Puype, Jan
1985 *Proceedings of the 1984 Trade Gun Conference.* Rochester, N.Y.: Rochester Museum and Science Center.

Queen Anne's Revenge Shipwreck Project
2011 *Rupert Method.* https//farm9.staticflickr.com/8295/7936892854_200a49df1a_b.jpg (accessed October 16, 2015).

Rohsenow, Warren M., and James P. Hartnett
1973 *Handbook of Heat Transfer.* New York: McGraw-Hill.

Rupert Harris Conservation
2013 "The Conservation of Lead Sculpture: Lead Dioxide Formation." www.rupertharris.com/final/sc_leads/examples/lead_dioxide/lead_dioxide_formation.php (accessed April 16, 2013).

Schmader, Matthew
2014 "The Slingstones and Arrows of Unfortunate Outrage: Vázquez de Coronado and the 'Tiguex War' of 1540–1542." Paper presented at the Eighth Biennial Meeting of Fields of Conflict International Battlefield Archaeology Conference, Columbia, S.C., March 11–16.

Scott, Douglas D.
2013 *Uncovering History: Archaeological Investigations at the Little Bighorn.* Norman: University of Oklahoma Press.

Scott, Douglas D., Richard A. Fox, Jr., and Melissa A. Connor
1989 *Archaeological Perspectives on the Battle of the Little Bighorn.* Norman: University of Oklahoma Press.

Scott, Douglas D., Richard A. Fox, Jr., and Dick Harmon
1987 *Archaeological Insights into the Custer Battle: An Assessment of the 1984 Field Season.* Norman: University of Oklahoma Press.

Simms, Jeptha R.
1882 *The Frontiersman of New York.* Vol. 1. Albany, N.Y.: G. C. Riggs.

Sivilich, Daniel
1996 "Analyzing Musket Balls to Interpret a Revolutionary War Site." *Historical Archaeology* 30 (2): 101–109.
2004 "Revolutionary War Musket Ball Typology—An Analysis of Lead Artifacts Excavated at Monmouth Battlefield State Park." Paper presented at American Battlefield Protection Program Seventh National Conference on Battlefield Preservation, Nashville, Tenn., April 19–25.
2005 "Evolution of Macro-Archaeology of the Battle of Monmouth—1778 American Revolutionary War." In *Battlefields Annual Review,* edited by Jon Cooksey, 72–85. South Yorkshire, U.K.: Pen and Sword Books.
2009 "What the Musket Ball Can Tell You: Monmouth Battlefield State Park, New Jersey." In *Fields of Conflict: Battlefield Archaeology from the Roman Empire to the Korean War,* edited by Douglas D. Scott, Lawrence Babits, and Charles Haecker, 84–101. Dulles, Va.: Potomac Books.
2011 "Analysis of Possible Military Artifacts Recovered from Partial Ship Remains Discovered at the Base of the World Trade Center Reconstruction Site, New York City, New York." Cultural resource report

REFERENCES

prepared for AKRF, Inc., Environmental, Planning, and Engineering Consultants, New York.

Sivilich, Daniel, and Garry Wheeler Stone
2009 "The Archaeology of Molly Pitcher." *American Revolution,* October 2009, 11–14.

Sivilich, Eric D., and Daniel M. Sivilich
2010 "Surveying, Statistics and Spatial Mapping: Predictive Modeling of 18th-Century Artillery at Monmouth Battlefield State Park, NJ." Paper presented at Council for Northeast Historical Archaeology annual meeting, Baltimore, Md.

Smith, George
1779 *An Universal Military Dictionary.* London: J. Millan.

Stark, Caleb, and John Stark
1860 *Memoir and Official Correspondence of Gen. John Stark, with Notices of Several Other Officers of the Revolution.* Concord, N.H.: G. Parker Lyon.

Stone, Garry Wheeler, Daniel Sivilich, and Mark Lender
1996 "A Deadly Minuet: The Advance of the New England 'Picked Men' against the Royal Highlanders at the Battle of Monmouth." *Brigade Dispatch,* 2–18.

Thacher, James
1823 *A Military Journal during the American Revolutionary War, from 1775 to 1783.* Boston, Mass.: Richardson and Lord.

Thomas, Bob
2011 "Why Are Gray Squirrels Eating Me Out of House and Lead?" Loyola University New Orleans, Loyola Center for Environmental Communication. http://loyno.edu/lucec/natural-history-writings/why-are-gray-squirrels-eating-me-out-house-lead (accessed October 26, 2013).

Thomas, Dean S.
1997 *Round Ball to Rimfire: A History of Civil War Small Arms Ammunition, Part One.* Gettysburg, Pa.: Thomas Publications.

Thompson, J. G.
1930 "Properties of Lead-Bismuth, Lead-Tin, Type Metal, and Fusible Alloys." *Bureau of Standards Journal of Research* 5 (5).

Topsell, Edward
[1607] *Of the Wilde Boar.* http://digital.lib.uh.edu/collection/p15195coll18/item/30 (accessed February 13, 2014).

von Steuben, Fredrick William
1985 *Baron von Steuben's Revolutionary War Drill Manual: A Facsimile Reprint of the 1794 Edition.* Mineola, N.Y.: Dover Publications.

Warfel, Stephen G.
2001 *Historical Archaeology at Ephrata Cloister: A Report on 2000 Investigations.* Edited by the State Museum of Pennsylvania. Harrisburg: Pennsylvania Historical and Museum Commission.

Washburn, Emory
1860 *Historical Sketches of the Town of Leicester, Massachusetts.* Boston, Mass.: John Wilson and Son. Copy provided by Jason Wickersty.

Washington, George
1775 "Proceedings of the Committee of Conference with General Washington, Oct. 22." In *American Archives: Documents of the American Revolution, 1774—1776,* vol. 3. amarch.lib.niu.edu/islandora/object/niu-amarch%3A106416.
1777–78 George Washington Papers at the Library of Congress, 1741–1799. Series 3b Varick Transcripts, General Orders. Available at memory.loc.gov.

Webster, Donovan
1999 "Pirates of the *Whydah.*" *National Geographic* 195 (5): 64–77.

Williams, Jonathon
1984 *De Scheel's Treatise on Artillery.* Bloomfield, Ont.: Museum Restoration Services.

Wolcott, Oliver, Sr.
1776 Papers. Box 1, folder 2, Correspondence, January–July 1776. Connecticut Historical Society, Hartford.

INDEX

Ahearn, Bill, 8–10, 14, 31, 90
air pockets, 24, 118
Alexander, William (Lord Stirling) (Maj. Gen.), 92, 94
alloy, 16, 25, 26, 76, 89, 98, 99, 101, 110, 116–22, 127, 131, 148, 152, 161
antimony, 49, 60, 100, 110, 151
arquebus, 6, 156, 157

Babits, Lawrence, 68
Banastre, Tarleton (Lt. Col.), 68
Battle, Dan, 18
Battle of Aughrim (Ireland), 83
Battle of Blue Licks, Ken., 69
Battle of Brandywine, Pa., 173
Battle of Bunker Hill, Mass., 109
Battle of Boyne (Ireland), 41
Battle of Cooch's Bridge, Del., 82
Battle of Cowpens, S.C., 68
Battle of Hubbarton, Vt., 173
Battle of Kunersdorf (Poland), 67, 99
Battle of Monmouth (Monmouth Courthouse), N.J., 10, 11, 20, 26, 28–30, 32, 51, 54, 57, 65, 68, 80, 97, 99, 173
Battle of Paoli, Pa., 173
Battle of Poltava (Poland), 44
Battle of Princeton, N.J., 9, 29, 173
Battle of Pułtusk (Poland), 37, 44, 77, 84

Battle of Saratoga, N.Y., 38
Battle of Talamanca (Spain), 84
Battle of Waterloo, 57, 58
Battle of Walloomsac, N.Y., 109
Battle of Zboriv (Ukraine), 5, 43, 101, 173
bayonet(s), 8–11, 15, 29, 68, 157
Bilby, Joseph G., and Katherine Bilby Jenkins, 47
Bliemeister method, 151
bones, human, 57–59
Boughton Hill Village site, N.Y., 45
bow, 4
Brinnel Hardness Number (BHN), 98, 122
buck and ball, 30–36
buck shot: cold swaged, 152; modern, 145–46; molds, 145; "pewter," 152
bullets, modern conical, 143–44
buttons, 77, 116, 119, 127, 138, 139

caliber (definition), 18
Calver, William, and Bolton, Reginald, 39, 73, 74, 77, 138, 139
Campillo, Xavier-Rubio, 84
canine teeth, 111, 114, 160, 166
canister shot (definition), 93
Capitaine du Chesnoy, Michel, 51, 54, 96
Carter, William, 45

INDEX

Caruana, Adrian, 92, 94, 96
case shot: definition, 93; mixed, 93
casting flaws, 87
chain shot, artillery, 78
Civil War, American, 8, 9, 33, 92, 100, 158
Civil War, English, 44, 64, 89, 162
Clinton, Henry (Lt. Gen.), 11, 109
Conflict Archaeology International Research Network (CAIRN), 83, 88
Coronado, Francisco Vásquez de, 84–85
crossbow, 4–6

Dann site, N.Y., 16, 42, 45, 46, 66, 146
De Scheel, Otto, 94–96
density, 4, 24–27, 117–18, 140
dentition marks, 102, 104, 111, 159–69

flint wraps, 90–91
Foard, Glenn, 5, 44, 83, 86, 88, 89, 95, 96, 98
Fort Hunter, Pa., 90
Fort Montgomery, N.Y., 21, 41, 54, 76, 87, 107, 111, 120, 173
Fort Niagara, N.Y., 93, 140
Fort Putnam, N.Y., 39, 114
Fort Stanwix, N.Y., 145
Forty-Second Regiment of Foot, 57, 62, 68, 75, 78, 80, 92, 94, 95, 113
fouling, 22, 23, 38, 49, 68, 146
fowling, 23, 66, 145
Freienstein Castle (Switzerland), 6
French and Indian Wars, 8, 28, 90
frizzen, 8

gaming pieces, dice, 20, 106, 119, 120, 128–34
gaming tokens, 135–38
gang mold, 17, 23, 66, 146

Geographic Information Systems (GIS), 173
George, King, III, 121–23, 125, 127
grapeshot (definition), 93
Greene, Nathaniel (Maj. Gen.), 96
ground penetrating radar (GPR), 173

hail-shot, 93, 96, 98, 99
hammer, musket, 8
handgonne, 6
Hartgen Archaeological Associates, 57
Historic St. Mary's City, Md., 46, 112
Hornady Manufacturing Company, 24
Hussite war, 5

impurities, 24, 27, 80, 118, 164, 166
incisors (teeth), 58, 103, 105–107, 110–11, 160, 165–66
injury, gunshot, 58

Jamestown, Va., 7, 20, 145, 167
jaw screw, 90, 91
Jefferson Patterson Park, Md., 104–105

Kinkor, Ken, 135
King George III, 121–23, 125, 127
Knarrstrom, Bo, 43, 69
Koenig, Armin, 6
Kovalski, Paul, Jr., 58
Kunersdorf Battlefield (Poland), 67, 99

Landskrona Battlefield (Sweden), 43, 60
Layfayette, Marquis de, 96
lead: mines, 116; pencil, 138, 139
Library of Congress, 121, 129, 130, 133, 158
Linck, Dana, 100, 101, 107, 111
Litchfield, Conn., 122
loading blocks (rifle), 68
Loyalists, British, 69, 116, 117

INDEX

Mandzy, Adrian, 5, 43–45, 69, 70, 101
Martin, Joseph Plumb (Pvt.), 57, 60, 75, 94, 95, 112, 113
Mary Rose (ship), 4
Maryland Archaeological Conservation Laboratory, 14, 119
matchlock, 3, 6, 7, 20, 44, 84, 156, 157
Mellen, Thomas, 108
Merchants' Shot Works, Md., 150
Miller, Henry, 102, 110, 112, 159–69
Mills, John, 49
mines, lead, 116
molars (teeth), 103, 109–12, 115, 160, 164–67
mold(s), 5, 16, 17, 20–23, 26, 27, 41, 44, 66, 111, 129, 142, 146, 147, 163
mold seams (definition), 17
Monmouth County Historical Association, N.J., 17
Morgan, Daniel (Brig. Gen.), 68
Mount Zion, Pa., 140
Muller, John, 24, 27
musket balls: "barrel band," 48, 49; cylindrical, 5, 78–86, 89; European, 41, 42; extended sprue, 42–46; halved and quartered, 73–76; melted, 141, 142; with nails, 76–78; pulled, 36–40, 60, 61; with ramrod marks, 36; rejected, 41; ricochet, 53, 54; with small cylindrical cavities, 86–88; surgically extracted, 59, 60. *See also* musket balls, chewed; musket balls, impacted
musket balls, chewed: by deer, 107, 108; by human, 108–115, 159–63, 166–69; by rodents, 105–107, 160, 164–66, 168
musket balls, impacted: with fence rail, 54, 55; human hit, 57–59; with musket barrel, 56, 57; with smooth, flat object, 55, 56; with soft target, 18, 48; with tree, 49–53
muskets: flintlock, 8, 36, 114, 117, 157, 158; fowling, 23, 66; infantry, 8, 28, 31; naval, 31; snaphaunce, 7, 8, 20, 69, 157; wheel lock, 7, 20
musket worm, 38–40, 137

Native Americans, 44, 69
New Jersey Historical Society, 92
New York Historical Society, 40, 121, 122, 139
New York State Archaeological Association, 45

Old Tennent Church, N.J., 59
Orr, David Gerald, 11

Parkman, Colin, 50
patch(ed), 38, 49, 66–71, 79
patina (definition), 17
percussion cap, 158
Phillips, Ralph, 86, 92
Phoenix Shot Tower, Md., 150
Piedras Marcadas Pueblo site, N.M., 84
pirate(s), 81, 82, 135
pistol, 7, 16, 17, 29, 31, 46, 53, 170
Pope's Fort, Md., 46, 112, 162–63
Prots, Ivanna, 70

Queen Anne's Revenge (ship), 149

Raritan Landing site, N.J., 173
Revolutionary War, American, 8–10, 14, 18, 21, 23, 26, 29, 30, 32, 33, 39, 41, 53, 68, 70, 73, 76–78, 82, 89, 92, 104, 108, 114, 116, 117, 121, 122, 128, 138, 145, 152, 159, 164–65
Rochester Museum, N.Y., 16, 45, 93

Royal Armouries, U.K., 139, 145
Royal College of Surgeons (Scotland), 57
Rupert method, 147–149, 151

Schmader, Matt, 84
Scott, Charles (Brig. Gen.), 56
Scott, Douglas D., 175
Seibert, Michael, 122
shrapnel, 100, 101
Shrapnel, Henry, 100
Shiels, Damian, 41, 83
shot tower, 150, 151
sling shot, 4
sinkers, fishing/net, 128, 140, 141
Sivilich Formula, 24, 27
slow match, 6
"sluggs." *See* musket balls, cylindrical
specific gravity, 24, 116
sprue, casting (definition), 17
Stark, John (Gen.), 109
statue, leadened, 121–27
St. Leonard Plantation, M.D., 104, 105
Stone, Garry Wheeler, 50, 159, 169, 175
Straube, Beverly (Bly), 7, 20

Thomas Pettus site, Va., 167
tin: alloy(s), 98, 117, 118, 121, 122, 127; density of, 117

toy(s), 128, 139, 140
touchhole, 6, 7

Valley Forge, Pa., 141
vent hole, 6
Von Steuben, Baron Friedrich Wilhelm (Gen.), 10, 29

Warfel, Stephen G., 140
Washington, George (Gen.), 11, 32, 58, 77, 128–30, 132
Washington Memorial Chapel, Pa., 11
Wayne, Anthony (Gen.), 54, 65, 96, 111, 173
weapons, edged, 63
weapons, psychological, 73, 76, 86
Webbs Landing site, Del., 163
Wheeler Village site, N.Y., 45
whizzers. *See* toy(s)
Whydah (ship), 81, 135, 137, 140
Williamites, 83
Wilson, Samuel, 64
windage, 18
Wixon, Scott, 122
World Trade Center ship, 20, 22, 88, 147, 149

x-ray fluorescence (XRF), 110, 111, 119, 120, 122, 123, 126, 127

zooarchaeologist, 102, 110

Cover art credits: (front top, left to right) "Pewter" alloy musket ball excavated at Monmouth Battlefield State Park, N.J. (photo by Daniel Sivilich); seventeenth-century ball excavated at the Pope's Fort site, Historic St. Mary's City, Md. (photo by Henry Miller, courtesy of Historic St. Mary's City); ball with square hole excavated at Pułtusk Battlefield, Poland (photo by Pawel Kobek, provided by Jakub Wrzosek of the National Heritage Board of Poland); eighteenth-century ball excavated at Monmouth Battlefield (photo by Daniel Sivilich); and ball with fabric wadding marks excavated at the site of the Battle of Kunersdorf, Poland (photo by Pawel Kobek, provided by Jakub Wrzosek of the National Heritage Board of Poland). (*front bottom*) Reenactor Eric Sivilich, June 2004, during the filming of the *Battlefield Detectives* episode on the Battle of Monmouth for the History Channel (photo by Lea Sivilich). (*back, left to right*) Ball hammered or rolled to create a slug, excavated at Pułtusk Battlefield, Poland (photo by Pawel Kobek, provided by Jakub Wrzosek of the National Heritage Board of Poland); canister shot excavated at the site of the Battle of Kunersdorf, Poland (photo by Pawel Kobek, provided by Jakub Wrzosek of the National Heritage Board of Poland); swine-chewed ball found at the Point of Woods at Monmouth Battlefield; and canister shot excavated at Monmouth Battlefield (photo by Daniel Sivilich).

Copyedited by Sally Bennett Boyington
Design and composition by Westchester Publishing Services
Set in Bembo and Trajan Pro
Cover design by Tony Roberts
Image prepress by University of Oklahoma Printing Services
Printed and bound by Versa Press, Inc.